SpringerBriefs in Food, Health, and Nutrition

Springer Briefs in Food, Health, and Nutrition present concise summaries of cutting edge research and practical applications across a wide range of topics related to the field of food science, including its impact and relationship to health and nutrition. Subjects include: Food Chemistry, including analytical methods; ingredient functionality; physic-chemical aspects; thermodynamics Food Microbiology, including food safety; fermentation; foodborne pathogens; detection methods Food Process Engineering, including unit operations; mass transfer; heating, chilling and freezing; thermal and non-thermal processing, new technologies Food Physics, including material science; rheology, chewing/mastication Food Policy And applications to: Sensory Science Packaging Food Qualtiy Product Development We are especially interested in how these areas impact or are related to health and nutrition. Featuring compact volumes of 50 to 125 pages, the series covers a range of content from professional to academic. Typical topics might include:

- A timely report of state-of-the art analytical techniques
- A bridge between new research results, as published in journal articles, and a contextual literature review
- A snapshot of a hot or emerging topic
- An in-depth case study
- A presentation of core concepts that students must understand in order to make independent contributions

Kalina Sikorska-Zimny

Sulforaphane in Human Health

Molecular Mechanisms, Pharmacokinetics, and Therapeutic Potential

Springer

Kalina Sikorska-Zimny
The National Institute of Horticultural Research
Skierniewice, Poland

ISSN 2197-571X ISSN 2197-5728 (electronic)
SpringerBriefs in Food, Health, and Nutrition
ISBN 978-3-032-34330-7 ISBN 978-3-032-34331-4 (eBook)
https://doi.org/10.1007/978-3-032-34331-4

This Springer imprint is published by the registered company Springer Nature Switzerland AG
The registered company address is: Gewerbestrasse 11, 6330 Cham, Switzerland

Preface

Sulforaphane has attracted scientific interest for decades, although its biological significance remained unclear for much of that time. A breakthrough came in 1992, when sulforaphane was identified as the principal chemopreventive compound in broccoli. This discovery transformed scientific perspectives on the role of diet in human health, shifting attention from cancer treatment towards cancer prevention and initiating decades of intensive research into the biological potential of cruciferous vegetables.

This book is, in many respects, the product of those three decades, filtered through 16 years of my own work at The National Institute of Horticultural Research, where the biochemistry of bioactive plant compounds—and sulforaphane in particular—has been the central focus of my scientific life. However, my path to this subject was not direct. My doctoral work in pharmacy and subsequent engineering degree from the Lodz University of Technology both centred on food science and food analysis, disciplines that trained me to think carefully about the gap between what a plant contains and what the human body ultimately receives. That gap, I would come to understand, is precisely where the most important questions about sulforaphane reside.

The compound itself is deceptively simple in structure yet remarkable in behaviour. It does not exist preformed in the plant. It is generated enzymatically—through the hydrolysis of its glucosinolate precursor, glucoraphanin, by the enzyme myrosinase—and this dependency on enzymatic conversion makes its chemistry sensitive to every variable imaginable: cultivar, growing conditions, processing temperature, cooking method, gut microbiota composition, and individual pharmacogenetics. Understanding sulforaphane therefore requires understanding an entire biological system, not merely a molecule.

This book attempts to provide that understanding in a structured and comprehensive way. It opens with the history of sulforaphane's discovery and traces the parallel history of myrosinase research, which began, remarkably, as early as 1831. It then examines the compound's pharmacokinetics and molecular mechanisms of action—including its well-characterised activation of the Nrf2/ARE pathway, its anti-inflammatory effects through NF-κB and NLRP3 inhibition, and its epigenetic activity through HDAC and DNMT modulation. Subsequent chapters survey the therapeutic evidence across oncology, neurology, cardiology, metabolic disease, and infectious disease, before turning to the clinical trial literature and the translational gap that separates preclinical promise from confirmed human benefit. The book concludes with an assessment of advanced formulation strategies—nanocarriers, cyclodextrin complexes, protein- and polysaccharide-based encapsulation systems—that represent the most credible current pathway towards a stable, bioavailable, clinically deployable sulforaphane-based therapeutic.

Throughout, I have tried to maintain a distinction that I consider essential in this field: the difference between effects observed at physiological concentrations attainable through diet or supplementation, and those demonstrated only at supraphysiological doses whose clinical relevance remains uncertain. This distinction matters enormously when translating laboratory findings into dietary or therapeutic recommendations, and I have endeavoured to apply it consistently.

This book is written for researchers, clinicians, and advanced students working at the intersection of phytochemistry, pharmacology, and nutritional science. I hope this volume serves that purpose.

My earlier work as co-author of a biochemistry textbook for undergraduate students taught me that clarity and precision are not in competition—that it is possible to be both scientifically exact and genuinely readable. I have tried to apply that lesson here.

The broccoli on your plate contains approximately the same glucoraphanin it always did. Whether your body converts it to sulforaphane, and in what quantity, depends on factors we are only beginning to understand fully. That uncertainty is not a reason for pessimism—it is the reason this research continues, and the reason this book exists.

Declaration of Competing Interests The author declares that there are no competing interests associated with this work. The research presented in this book was conducted in the absence of any commercial or financial relationships that could be construed as a potential conflict of interest. No funding was received from industry sources, and the author has no financial stake in any product, company, or organisation related to the subject matter of this publication.

Skierniewice, Poland Kalina Sikorska-Zimny

Competing Interests The author has no competing interests to declare that are relevant to the content of this manuscript.

Contents

Chapter 1
Sulforaphane: Discovery, Chemistry, Biosynthesis, and Plant Biology

Abstract Sulforaphane (SFN) is a potent bioactive isothiocyanate derived primarily from cruciferous vegetables, with a research history spanning over a century. Initial observations date back to the late nineteenth century, with chemical synthesis of the compound achieved in 1948. A pivotal milestone came in 1992, when the anticancer properties of SFN from broccoli were discovered, shifting the global research focus from cancer treatment towards cancer prevention. SFN formation is closely linked to its glucosinolate precursor, glucoraphanin (GRN), whose hydrolysis catalysed by the enzyme myrosinase constitutes the central biochemical event underlying SFN production. Research on myrosinase dates back to its first observation in 1831, with further characterisations in 1840 and 1863. Myrosinase activity extends beyond plants, having been identified in phylogenetically diverse organisms including aphids, bacteria, and fungi, pointing to a broader ecological role in glucosinolate metabolism. The chemical factors governing the product spectrum of myrosinase-catalysed glucosinolate hydrolysis are also of central relevance, contributing to the biochemical variability of SFN production across biological systems and processing conditions, and forming the basis for understanding its subsequent pharmacological and clinical significance.

Keywords Sulforaphane · Glucoraphanin · Myrosinase · Glucosinolate hydrolysis

1.1 Historical Discovery and Chemical Properties of Sulforaphane

Sulforaphane (SFN) is a naturally occurring bioactive compound belonging to the isothiocyanate family, derived from glucosinolates primarily found in plants of the Brassicaceae family (Fig. 1.1). Chemically identified as 1-isothiocyanato-4-(methylsulfinyl)butane (EPA n.d.), it is a potent bioactive phytochemical found in cruciferous vegetables such as broccoli, Brussels sprouts, cabbage, cauliflower, kale, radish, and arugula (Balenović et al. 1966; Fahey and Talalay 1999; Kamal et al. 2020). In 1963, Kjaer established that these isothiocyanate-producing

K. Sikorska-Zimny, *Sulforaphane in Human Health*, SpringerBriefs in Food, Health, and Nutrition, https://doi.org/10.1007/978-3-032-34331-4_1

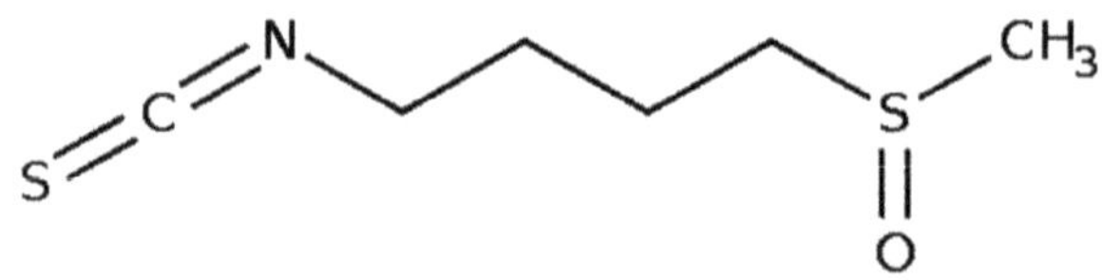

Fig. 1.1 Sulforaphane chemical formula (EPA n.d.)

glucosides are widely distributed in plant families such as Brassicaceae (formerly Cruciferae), Resedaceae, and Capparaceae.

In plants, sulforaphane acts as a natural chemical repellent, and in humans, it exhibits a wide range of health-promoting properties, making it one of the leading compounds studied in terms of shaping human health.

The scientific history of sulforaphane dates back to 1897, when Gadamer first proposed the glucosinolate structure (Gadamer 1897). In the mid-twentieth century, research on the antimicrobial properties of cruciferous plants has intensified. In 1947, Ivanovics and Horváth investigated radish extract (*Raphanus sativus* L.) and identified a compound termed "raphanin", which effectively inhibited the growth of *Staphylococcus* and *Bacterium coli* (Ivanovics and Horvath 1947). A year later, Schmid and Karrer isolated "sulforaphene", a structural analogue of sulforaphane, from radish seeds (Schmid and Karrer 1948b). In the same year, these researchers successfully synthesised a dihydroderivative of sulforaphane and characterised both D,L-sulforaphane enantiomers (Schmid and Karrer 1948b). The synthesised L-sulforaphane was chemically identical to the naturally occurring compound found in radish seeds (Schmid and Karrer 1948a).

A significant milestone was achieved in 1959 when Procházka, Šanda, and Jiroušek identified sulforaphane in plants of the *Brassica* genus (Procházka et al. 1959). That same year, Procházka successfully isolated sulforaphane from hoary cress (*Lepidium draba* L.), buckhorn plantain (*Plantago lanceolata* L.) (Procházka 1959a, b), savoy (*Brassica oleracea* var. *sabauda* L.), and red cabbage (*Brassica oleracea* var. *capitata* L.), highlighting its significant antimicrobial activity (Procházka 1959a, b; Fahey and Talalay 1999).

The modern pharmaceutical significance of sulforaphane was established in 1992 when Professors Talalay and Zhang were the first to isolate SFN from broccoli and demonstrate its anticarcinogenic effects (Zhang et al. 1992; Fahey and Talalay 1999). These discoveries shifted the direction of research from cancer treatment to cancer prevention (Johns Hopkins Medicine 2019).

These studies have highlighted the ability of plant extracts to induce carcinogen detoxification and chemoprotective enzyme activity (Fahey and Kensler 2021). These investigations also revealed a critical biochemical detail: plants contain almost no free sulforaphane. Instead, they store a stable glucosidic precursor known as glucoraphanin (Yagishita et al. 2019; Fahey and Kensler 2021).

Glucoraphanin is present in the highest concentrations within the seeds and is predominantly stored in specialised S-cells located between the endodermis and the phloem within the vascular bundle. This localisation was confirmed in 2000 through dispersive X-ray imaging by Koroleva et al. (2000). The conversion of inactive glucoraphanin to the active isothiocyanate sulforaphane requires hydrolysis catalysed by the enzyme myrosinase (Mangla et al. 2021). This biochemical transformation

Table 1.1 Chemical properties of sulforaphane (EPA n.d.; NLM-NCBI 2025; HMDB 0005792, n.d.; Phytohub, n.d.; Chemspider, n.d.)

Chemical name	1-isothiocyanato-4-methanesulfinylbutane 4-(methylsulfinyl)butyl isothiocyanate
Synonyms	Broccoli sprouts extract DL-sulforaphane Sulforafan Sulforaphane Sulphoraphane
Boiling point	368–369 °C @ 760.00 mmHg
Melting point	Not available
Water solubility	13,820 mg/L @ 25 °C
Chemical form	Powder
CAS no.	4478-93-7
European Community (EC) number	878-934-0
Molecular formula	C6H11NOS2
Exact mass	177.02820632 g/mol

typically occurs upon the structural injury of the plant tissue, such as chewing or chopping, or via the action of human gut microbiota (Lagunas-Rangel 2024; Sikorska-Zimny and Beneduce 2021).

Over the last three decades, SFN has gained significant attention in pharmaceutical research because of its multifaceted biological activities, including antioxidant, anti-inflammatory, neuroprotective, and potent anticancer properties (Mangla et al. 2021). In humans, it supports cardiovascular and hepatic health, contributes to the management of diabetes and obesity, and is currently being investigated for potential therapeutic applications in the treatment of autism spectrum disorder.

Chemically, sulforaphane is both a sulfoxide and an isothiocyanate, characterised by its light-yellow colouration and molecular weight of 177.3 g/mol (PubChem n.d.; Bansal et al. 2018) (Table 1.1).

However, SFN is inherently unstable because it is highly sensitive to light and heat and undergoes rapid oxidation (Wu et al. 2010). Consequently, its formulation is critically dependent on factors such as pH, buffer systems, and solvent composition (Wu et al. 2014) (Fig. 1.2). The Chemical Entities of Biological Interest (ChEBI) database classifies sulforaphane as an "antineoplastic agent, antioxidant, histone deacetylase inhibitor, plant metabolite, isothiocyanate and sulfoxide" (ChEBI n.d.).

1.2 Glucoraphanin Biosynthesis Pathways and Myrosinase System

The precursor of SFN is glucoraphanin (GRA), the biosynthesis of which involves a series of enzymatic steps that have been extensively characterised. The detailed mechanism, including the genetic components, has been elucidated in *Arabidopsis thaliana* by Yin et al. (2015) and Mikkelsen et al. (2010).

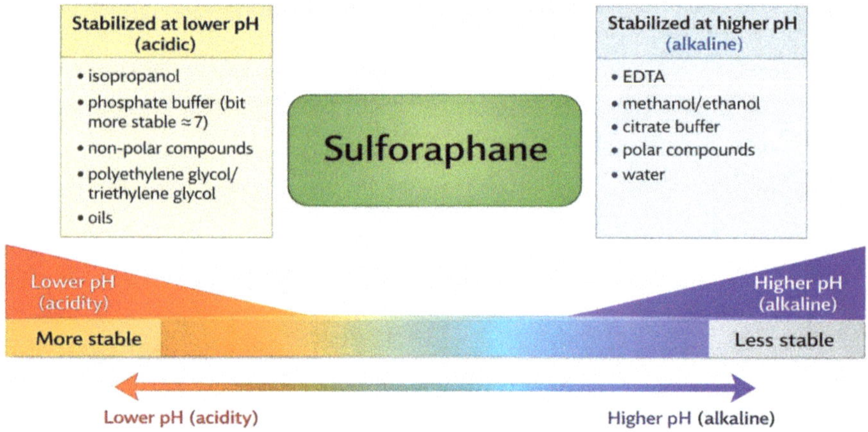

Fig. 1.2 Factors influencing the stability of sulforaphane (Franklin et al. 2014; Zhang et al. 2020; Yuanfeng et al. 2021)

The initial step occurs in the chloroplast, where methionine undergoes cytosolic transamination catalysed by branched-chain amino transferase 4 (BCAT4, operates at the chloroplast envelope), yielding 4-methylthio-2-oxobutanoic acid (Fig. 1.3). This intermediate is subsequently elongated by the sequential addition of two methylene groups in a process involving three rounds of transformation. Elongation begins with the condensation of acetyl-CoA, a reaction catalysed by methylthioalkyl malate synthase (MAM), followed by isomerisation mediated by isopropylmalate isomerase (IPMI), and concluded with oxidative decarboxylation facilitated by isopropylmalate dehydrogenase (IPMDH). The resulting product, 7-(methylthio)-2-oxoheptanoic acid, is converted into dihomomethionine (6-(methylthio)hexylhomoserine or more precisely L-dihomomethionine) via transamination by BCAT 3 (Hansen et al. 2001; Yin et al. 2015).

Subsequent steps leading to the formation of the core glucosinolate structure occur in the cytosol with enzymes (present in the cytosol and ER/vacuole membrane). Here, cytochrome P450 enzymes of the CYP79 family catalyse the conversion of amino acids into aldoximes, which are further processed by CYP83 into S-glutathionyl-thiohydroximate (Hansen et al. 2001; Yi et al. 2015; Yin et al. 2015). S-Glutathionyl-thiohydroximate undergoes sequential removal of the glutathione peptide chain through three enzymatic steps. First, GGP1 (gamma-glutamyl peptidase), located at the cytosol/vacuolar membrane, cleaves the γ-glutamyl residue from the glutathione moiety, releasing free glutamate and yielding S-cysteinylglycyl-thiohydroximate. Second, CN peptidase (cysteinylglycine dipeptidase) cleaves the remaining Cys-Gly dipeptide bond, releasing free glycine and yielding S-cysteinyl-thiohydroximate. Third, SUR1 (C-S lyase) cleaves the carbon–sulphur bond between cysteine and thiohydroximate carbon, releasing pyruvate and ammonia and yielding a free thiohydroximate-peptide-free intermediate in the pathway. UGT74C1 (UDP-glucosyltransferase), which acts in the cytosol, transfers glucose from

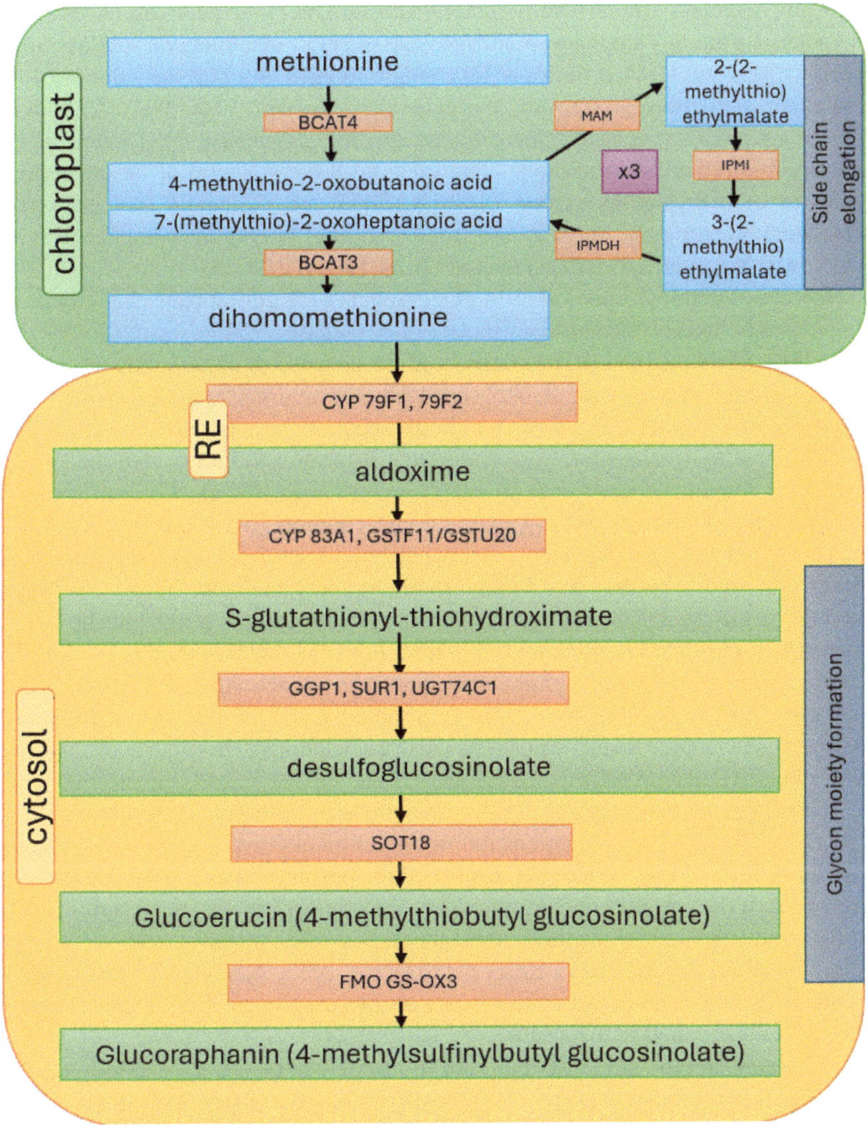

Fig. 1.3 Biosynthesis of glucoraphanin (Geu-Flores et al. 2009; Bednarek 2012; Yin et al. 2015; Crocoll et al. 2016; Yagishita et al. 2019; Nguyen et al. 2020; Sikorska-Zimny and Beneduce 2021; Zhao et al. 2025)

UDP-glucose to the free thiol group of the thiohydroximate, forming the S-glucoside bond characteristic of all glucosinolates and releasing UDP, yielding desulfoglucosinolate. In the next step, a sulphate group from the PAPS (3′-phosphoadenosine-5′-phosphosulfate) donor is transferred onto the hydroxyl group of the oxime nitrogen by SOT18 (sulfotransferase). This sulfation step yields glucoerucin (4-(methylthio) butyl glucosinolate), the first fully intact glucosinolate in the pathway, carrying the characteristic anionic sulphate on nitrogen and the glucose moiety on sulphur. Since erucin (ERU) and GRA differ only by a single oxygen atom on the sulphur of the side chain methylthio group, only oxygenation is needed. FMO GS-OX3 (flavin-containing monooxygenase, glucosinolate-specific oxygenase 3), located at the endoplasmic reticulum membrane, catalyses the S-oxygenation of the methylthio group to a methylsulfinyl group (with NADPH and molecular oxygen as co-substrates), yielding glucoraphanin (4-(methylsulfinyl)butyl glucosinolate) (Hansen et al. 2001; Yin et al. 2015; Zhuang et al. 2021; Liu et al. 2023).

However, the transformation from glucoraphanin to sulforaphane can only occur via enzymatic hydrolysis in the presence of myrosinase.

In 1831, Robiquet and Boutron observed the absence of volatile oils when mustard seeds were treated with alcohol, hot water, or acidified water. This led to the hypothesis that certain compounds within the seeds are sensitive to these conditions and are prone to degradation. Nine years later, Bourton and Frémy identified a substance that coagulated in water at 70–80 °C, as well as in 40% alcohol, sulfuric acid, or potassium solutions, with irreversible transformations. They attributed this substance—referred to as "emulsion"—to the enzymatic conversion of sinapsin into a "bitter principle" (Boutron and Fremy 1840). In the same year, Bussy (1840) isolated two distinct substances resulting from the reaction with water, which he named *Myronsäure* (*acide myronique*/myron essence/myronic acid) and *myrosyne*, where "*myron essence*" and "*syn*" signified their interdependence. Although Bussy did not fully elucidate the mechanism of myrosinase activity, as he was unable to crystallise the enzyme in its pure form, he noted its close association with amino acids, particularly albumin (Bussy 1840).

The enzymatic degradation of glucosinolates was further described by Will and Körner (1863). Their investigations into the reaction of "myron acid" under various conditions, including different pH levels and the presence of heavy metals, demonstrated that it decomposes under myrosinase into mustard oil, dextrose, sulfuric acid (some conditions), other insoluble substances, possible cyanoallil, and free sulphur (some conditions) (Will and Körner 1863). In 1933, Neuberg and Schonbeck pointed out the possibility of the existence of different types of myrosinase or sinigraze, as it was called in the article.

At present, myrosinase (EC 3.2.1.147), a member of the glycoside hydrolase family 1, is an enzyme that specifically hydrolyses thio-linked glycosides, a function known exclusively in plants. It has a molecular weight of approximately 157 kDa, with isozymic subunits ranging between 40 and 70 kDa, depending on the plant species (Bones and Rossiter 2006; Mahn et al. 2014; Huang et al. 2023). Mahn et al. (2014) further highlighted the presence of distinct myrosinase isoforms exhibiting substrate specificity across different plant species (Eriksson et al. 2001; Mahn et al. 2014).

Several types of myrosinases have been identified. The most extensively characterised MYR is MYR I, comprising the MA, MB, and MC isoforms in *Brassica napus*. Similarly, MYR II, encompassing the AtTGG1–AtTGG6 subfamilies in *Arabidopsis thaliana*, has been well documented (Nakano et al. 2017). Notably,

Arabidopsis thaliana possesses two major myrosinases, TGG1 and TGG2, which are stored in specialised myrosin cells (Kissen et al. 2009; Restović et al. 2024). Additionally, atypical myrosinases, such as PENETRATION 2 (PEN2) and PYK10, were identified by Bednarek et al. (2009) and Nakano et al. (2017), respectively. The distribution of myrosinase isoforms appears to correlate with plant morphology. Angelino et al. (2015) reported that MYR I isoforms are predominantly localised in aerial tissues, whereas MYR II isoforms are primarily found in roots (Andersson et al. 2009; Kissen et al. 2009; Angelino et al. 2015).

The precise origin of myrosin cells is still unclear. It has been postulated that ancestral *Brassicales* accumulated glucosinolates within stomatal guard cells, where they were hydrolysed into isothiocyanates and secreted into the extracellular space via transport proteins such as PEN3. This mechanism likely evolved as a defence strategy against bacterial invasion through stomatal openings (Bednarek 2012; Shirakawa and Hara-Nishimura 2018). Subsequent studies have confirmed that myrosinase-containing idioblasts, known as myrosin cells, are primarily localised in the phloem tissue, where they are positioned peripherally (Restović et al. 2024). These cells differ significantly in size and content from the surrounding cells and are highly enriched in proteins. Furthermore, their morphology and distribution vary depending on the plant organ in which they occur and the developmental stage of the plant (Andréasson et al. 2001; Kissen et al. 2009; Shirakawa and Hara-Nishimura 2018). Myrosinase activity also varies substantially depending on its source. Sangkret et al. (2019) quantified enzymatic activity at 0.1394 U/g dry mass in mustard seeds and 1.1470 U/g in broccoli (Sangkret et al. 2019).

Beyond plants, myrosinase has been identified in fungi, including *Aspergillus sydowi*, *Aspergillus niger*, and *Geotrichum candidum*. Smits et al. (1993) demonstrated that *Aspergillus clavatus II-9* and *Fusarium oxysporum @ 149* can degrade glucosinolates into glucose and allyl isothiocyanate, suggesting their potential role in detoxifying mustard seeds for animal feed (Smits et al. 1993; Abdel-Massih et al. 2023). Additionally, myrosinase activity has been documented in aphids such as *Brevicoryne brassicae* (Jones et al. 2001; Francis et al. 2002).

Several bacterial species also exhibit myrosinase activity, including *Enterococcus casseliflavus*, *Bacteroides thetaiotaomicron*, and members of the genera *Bacillus*, *Citrobacter*, *Companilactobacillus*, *Enterobacter*, *Enterococcus*, *Lactobacillus*, *Lactococcus*, and *Leclercia* (Yuanfeng et al. 2021; Youseif et al. 2022; Abdel-Massih et al. 2023). However, some bacterial species, such as *Agrobacterium tumefaciens* and *Sarcina lutea*, degrade glucosinolates through mechanisms other than hydrolysis (Smits et al. 1993).

Nakano et al. (2017) reported significant structural and functional differences between myrosinases across different biological kingdoms. For instance, insect myrosinases use distinct glutamate residues as proton donors, whereas bacterial myrosinases belong to glycosyl hydrolase family 3 and diverge from their plant counterparts (Nakano et al. 2017). These findings suggest that myrosinases have evolved independently in different lineages (Shirakawa and Hara-Nishimura 2018). Fungal myrosinases also exhibit notable differences from plant enzymes, particularly in their insensitivity to ascorbic acid and distinct affinities for specific glucosinolates (Ohtsuru et al. 1969). Furthermore, aphid myrosinases have been found to differ from plant-derived enzymes in terms of substrate specificity and glucosinolate-binding affinities (Jones et al. 2001).

Myrosinase is highly sensitive to heat, with inactivation occurring at temperatures exceeding 40 °C, depending on the plant source, including cultivar and organ specificity of the plant. Most enzymes are estimated to become inactivated at temperatures above 60 °C (Yuanfeng et al. 2021). The activity of myrosinase is modulated by various external factors, such as ascorbic acid and zinc ions (Liang et al. 2006; Bhat and Vyas 2019). The optimal temperature range for plant-derived myrosinase activity is between 20 and 70 °C, with an effective pH range of 4–8 (Yuanfeng et al. 2021).

Myrosinase plays a pivotal role in plant defence mechanisms. Upon tissue disruption caused by pathogens, insects, or herbivores, the enzyme comes into contact with glucosinolates (GLS), catalysing their hydrolysis into biologically active derivatives, including isothiocyanates, thiocyanates, and nitriles, ranked by their potency (Halkier and Gershenzon 2006; Gimsing and Kirkegaard 2009; Sikorska-Zimny and Beneduce 2021). This biochemical defence strategy is commonly referred to as the "mustard oil bomb" theory (Kissen et al. 2009; Restović et al. 2024). As previously noted, given the extracellular localisation of GLS relative to the phloem, this system appears to function as an effective long-distance transport and defence mechanism (Koroleva et al. 2000).

Yagishita et al. (2019) suggested that glucosinolates may play a broader role in plant adaptation to environmental stressors, as evidenced by the higher GLS concentrations observed in wild-type varieties than in their cultivated counterparts (Yagishita et al. 2019). Intriguingly, this plant defence mechanism appears to have been co-opted by certain herbivorous insects as a means of self-protection. Bridges et al. (2002) investigated two aphid species, *Brevicoryne brassicae* (cabbage aphid) and *Lipaphis erysimi* (turnip aphid) and reported that these insects can sequester glucosinolates from their host plants without converting them into toxic metabolites. Upon predator attack, these aphids can enzymatically release isothiocyanates, which function similarly to the alarm pheromone *E*-β-farnesene, thereby providing chemical defence against predators (Bridges et al. 2002).

Different insect species have evolved distinct strategies to cope with glucosinolate-based plant defences. For instance, the small white butterfly (*Pieris rapae*) employs nitrile-specifier proteins to modify the hydrolysis products of glucosinolates, whereas other insects, such as *Brevicoryne brassicae*, tolerate and actively sequester these compounds for their own defence (Kissen et al. 2009). A distinct detoxification strategy is employed by the diamondback moth (*Plutella xylostella*), which produces glucosinolate sulfatase (GSS), an enzyme that converts glucosinolates into desulfo-glucosinolates, which cannot be hydrolysed by myrosinase, thereby neutralising the plant's chemical defence system (Ratzka et al. 2002).

Shirakawa and Hara-Nishimura (2018) proposed that these sophisticated counteradaptations are the result of a long-standing evolutionary "*an arms race between Brassicales plants and their pests and pathogens*". This dynamic interplay between plant defence and insect counter-defence highlights the complexity of coevolutionary processes in ecological systems (Shirakawa and Hara-Nishimura 2018).

The myrosinase enzyme facilitates the hydrolysis of the β-thioglucoside bond, yielding aglycones that subsequently undergo an intramolecular rearrangement of hydroxamic acids to isothiocyanates, as described in the Lossen rearrangement (Lossen 1865; Fahey and Talalay 1999; Matsugi and Shioiri 2023). The formation of sulforaphane is favoured under neutral to slightly acidic conditions (pH 6–7). Yuanfeng et al. (2021) further suggested that the reaction can occur effectively within a broader pH range of 5–8. Ares et al. (2014) estimated that the conversion of glucoraphanin to sulforaphane in broccoli occurs over a period of approximately 2–4 h.

In addition, the presence of certain compounds influences the formation of specific compounds. Sulforaphane nitrile is facilitated in the presence of ferrous ions and epithiospecifier protein (ESP), both of which function as non-catalytic cofactors of myrosinase and are highly heat-sensitive (Matusheski et al. 2004; Backenköhler et al. 2018; Yuanfeng et al. 2021). The role of ESP in this process was confirmed by de Torres Zabala et al. (2005) in *Arabidopsis thaliana*, where overexpression of ESP led to increased production of nitriles. ESP mediates the breakdown of the thiohydroximate O-sulfinate intermediate, thereby reducing the sulforaphane yield (Matusheski et al. 2004). Consequently, to maximise sulforaphane production, it is essential to maintain a specific thermal threshold that deactivates ESP while preserving the myrosinase activity.

References

Abdel-Massih RM, Debs E, Othman L, Attieh J, Cabrerizo FM (2023) Glucosinolates, a natural chemical arsenal: more to tell than the myrosinase story. Front Microbiol 14:1130208

Andersson D, Chakrabarty R, Bejai S, Zhang J, Rask L, Meijer J (2009) Myrosinases from roots and leaves of *Arabidopsis thaliana* differ in catalytic properties. Phytochemistry 70:1345–1354

Andréasson E, Jørgensen LB, Höglund AS, Rask L, Meijer J (2001) Different myrosinase and idioblast distribution in *Arabidopsis* and *Brassica napus*. Plant Physiol 127:1750–1763

Angelino D, Dosz EB, Sun J, Hoeflinger JL, Van Tassell ML, Chen P et al (2015) Myrosinase-dependent and independent formation of isothiocyanates. Front Plant Sci 6:831

Ares A, Bernal J, Martín MT, Bernal JL, Nozal MJ (2014) Optimized formation and determination of sulforaphane in broccoli. Food Anal Methods 7:730–740

Backenköhler A, Eisenschmidt D, Schneegans N, Strieker M, Brandt W, Wittstock U (2018) Iron as cofactor of specifier proteins in glucosinolate breakdown. PLoS One 13:e0205755

Balenović K, Deljac A, Monković I, Štefanac Z (1966) Synthesis of (±) sulphoraphene. Tetrahedron 22:2139–2143

Bansal M, Singh N, Pal S, Dev I, Ansari KM (2018) Chemopreventive role of phytochemicals in colorectal cancer. Adv Mol Toxicol 12:69–121

Bednarek P (2012) Sulfur-containing secondary metabolites in plant immunity. ChemBioChem 13:1846–1859

Bednarek P, Pislewska-Bednarek M, Svatos A, Schneider B, Doubsky J, Mansurova M (2009) Glucosinolate metabolism mediates antifungal defense. Science 323:101–106

Bhat R, Vyas D (2019) Myrosinase: structural and catalytic insights. Crit Rev Biotechnol 39:508–523

Bones AM, Rossiter JT (2006) Enzymic and chemical decomposition of glucosinolates. Phytochemistry 67:1053–1067

Boutron F, Fremy E (1840) Untersuchung des schwarzen und weissen Senssamens. Justus Liebigs Ann Chem 34:230–232

Bridges M, Jones AM, Bones AM, Hodgson C, Cole R, Bartlet E (2002) Spatial organization of the glucosinolate-myrosinase system in brassica specialist aphids is similar to that of the host plant. Proc R Soc Lond B 269(1487):187–191. https://doi.org/10.1098/rspb.2001.1861

Bussy A (1840) Untersuchungen über die Bildung des ätherischen Senföls. Justus Liebigs Ann Chem 34(2):223–230. https://doi.org/10.1002/jlac.18400340211

Chemical Entities of Biological Interest (ChEBI) (n.d.) 47807 – sulforaphane. https://www.ebi.ac.uk/chebi/searchId.do?chebiId=CHEBI:47807. Accessed 17 Feb 2025

Chemspider (n.d.) Chemical structure 7827557. https://www.chemspider.com/Chemical-Structure.7827557.html. Accessed 4 Nov 2025

Crocoll C, Mirza N, Reichelt M, Gershenzon J, Halkier BA (2016) Optimization of engineered production of the glucoraphanin precursor dihomomethionine in *Nicotiana benthamiana*. Front Bioeng Biotechnol 4:14

de Torres Zabala M, Grant M, Bones AM, Bennett R, Lim YS, Kissen R, Rossiter JT (2005) Characterisation of recombinant epithiospecifier protein and its over-expression in *Arabidopsis thaliana*. Phytochemistry 66(8):859–867

Environmental Protection Agency (n.d.) Sulforaphane, 4478-93-7 | DTXSID8036732. https://comptox.epa.gov/dashboard/chemical/synonyms/DTXSID8036732. Accessed 17 Feb 2025

Eriksson S, Ek B, Xue J, Rask L, Meijer J (2001) Identification and characterization of soluble and insoluble myrosinase isoenzymes in different organs of *Sinapis alba*. Physiol Plant 111(3):353–364

Fahey JW, Kensler TW (2021) The challenges of designing and implementing clinical trials with broccoli sprouts… and turning evidence into public health action. Front Nutr 8:648788

Fahey JW, Talalay P (1999) Antioxidant functions of sulforaphane: a potent inducer of phase II detoxication enzymes. Food Chem Toxicol 37:973–979. https://doi.org/10.1016/S0278-6915(99)00082-4

Francis F, Lognay G, Wathelet JP, Haubruge E (2002) Characterisation of aphid myrosinase and degradation studies of glucosinolates. Arch Insect Biochem Physiol 50(4):173–182

Franklin SJ, Dickinson SE, Karlage KL, Bowden GT, Myrdal PB (2014) Stability of sulforaphane for topical formulation. Drug Dev Ind Pharm 40(4):494–502

Gadamer J (1897) Ueber das sinigrin. Ber Dtsch Chem Ges 30(2):2322–2327

Geu-Flores F, Nielsen MT, Nafisi M, Møldrup ME, Olsen CE, Motawia MS, Halkier BA (2009) Glucosinolate engineering identifies a γ-glutamyl peptidase. Nat Chem Biol 5(8):575–577

Gimsing AL, Kirkegaard JA (2009) Glucosinolates and biofumigation: fate of glucosinolates and their hydrolysis products in soil. Phytochem Rev 8:299–310

Halkier BA, Gershenzon J (2006) Biology and biochemistry of glucosinolates. Annu Rev Plant Biol 57:303–333. https://doi.org/10.1146/annurev.arplant.57.032905.105228

Hansen CH, Wittstock U, Olsen CE, Hick AJ, Pickett JA, Halkier BA (2001) Cytochrome P450 CYP79F1 from *Arabidopsis* catalyzes the conversion of dihomomethionine and trihomomethionine to the corresponding aldoximes in the biosynthesis of aliphatic glucosinolates. J Biol Chem 276(14):11078–11085

Huang Z, Liu N, Fang Y, Hou X, Yang G, Lu J, Liu S (2023) Purification and characterization of a cold-active myrosinase from marine *Pseudomonas oleovorans* SuMy07. Acta Biochim Biophys Sin 55(4):695

Ivanovics G, Horvath S (1947) Raphanin, an antibacterial principle of the radish (Raphanus sativus). Nature 160(4061):297

Johns Hopkins Medicine (2019) Paul Talalay, Hopkins researcher famed for finding cancer-preventing compound in broccoli, dies. https://www.hopkinsmedicine.org/news/newsroom/

news-releases/2019/03/paul-talalay-hopkins-researcher-famed-for-finding-cancer-preventing-compound-in-broccoli-dies. Accessed 17 Feb 2025

Jones AME, Bridges M, Bones AM, Cole R, Rossiter JT (2001) Purification and characterisation of a non-plant myrosinase from the cabbage aphid *Brevicoryne brassicae* (L.). Insect Biochem Mol Biol 31(1):1–5

Kamal MM, Akter S, Lin CN, Nazzal S (2020) Sulforaphane as an anticancer molecule: mechanisms of action, synergistic effects, enhancement of drug safety, and delivery systems. Arch Pharm Res 43(4):371–384

Kissen R, Rossiter JT, Bones AM (2009) The 'mustard oil bomb': not so easy to assemble?! Localization, expression and distribution of the components of the myrosinase enzyme system. Phytochem Rev 8:69–86

Kjaer A (1963) Isothiocyanates of natural derivation. Pure Appl Chem 7(2–3):229–246

Koroleva OA, Davies A, Deeken R, Thorpe MR, Tomos AD, Hedrich R (2000) Identification of a new glucosinolate-rich cell type in Arabidopsis flower stalk. Plant Physiol 124(2):599–608

Lagunas-Rangel FA (2024) Sulforaphane target protein prediction: a bioinformatics analysis. Appl Sci 14(3):1052

Liang H, Yuan Q, Xiao Q (2006) Effects of metal ions on myrosinase activity and the formation of sulforaphane in broccoli seed. J Mol Catal B Enzym 43(1–4):19–22

Liu G, He H, Wang P, Zhao X, Ren F (2023) Glucoraphanin accumulation via glucoraphanin synthesis promotion during broccoli germination. Foods 13(1):41. https://doi.org/10.3390/foods13010041

Lossen W (1865) Uber das hydroxylamine. Z Chem 8:551–553

Mahn A, Angulo A, Cabanas F (2014) Purification and characterization of broccoli (*Brassica oleracea* var. italica) myrosinase (β-thioglucosidase glucohydrolase). J Agric Food Chem 62(48):11666–11671

Mangla B, Javed S, Sultan MH, Kumar P, Kohli K, Najmi A, Ahsan W (2021) Sulforaphane: a review of its therapeutic potentials, advances in its nanodelivery, recent patents, and clinical trials. Phytother Res 35(10):5440–5458

Matsugi M, Shioiri T (2023) Recent new developments in Hofmann, Curtius, Schmidt, Lossen, and related reactions. In: Reference module in chemistry, molecular sciences and chemical engineering. Elsevier. ISBN 978-0-12-409547-2

Matusheski NV, Juvik JA, Jeffery EH (2004) Heating decreases epithiospecifier protein activity and increases sulforaphane formation in broccoli. Phytochemistry 65(9):1273–1281

Mikkelsen MD, Olsen CE, Halkier BA (2010) Production of the cancer-preventive glucoraphanin in tobacco. Mol Plant 3:751–759

Nakano RT, Piślewska-Bednarek M, Yamada K, Edger PP, Miyahara M, Kondo M (2017) PYK10 myrosinase reveals a functional coordination between endoplasmic reticulum bodies and glucosinolates in *Arabidopsis thaliana*. Plant J 89:204–220

National Library of Medicine – National Center for Biotechnology Information (2025) Sulforaphane. https://pubchem.ncbi.nlm.nih.gov/compound/5350. Accessed 17 Feb 2025

Neuberg C, Schoenebeck O (1933) Über Aufteilung der Myrosinase. Naturwissenschaften 21:404–405. https://doi.org/10.1007/BF00599308 [uwaga: w oryginale błędny DOI — do weryfikacji]

Nguyen VT, Stewart J, Lopez M, Ioannou I, Allais F (2020) Glucosinolates: natural occurrence, biosynthesis, accessibility, isolation, structures, and biological activities. Molecules 25(19):4537

Ohtsuru M, Tsuruo I, Hata T (1969) Studies on fungous myrosinase: part I–III. Agric Biol Chem 33(9):1309–1325

Phytohub (n.d.) Sulforaphane PHUB000821. https://phytohub.eu/entries/PHUB000821. Accessed 21 Jan 2025

Procházka Ž (1959a) Chromatographic proof of the presence of sulforaphene in plantain. Naturwissenschaften 46:426. https://doi.org/10.1007/BF00599308

Procházka Ž (1959b) Isolation of sulforaphane from hoary cress (Lepidium draba L.). Collect Czechoslov Chem Commun 24(7):2429–2430

Procházka Z, Šanda V, Jirousek L (1959) Isothiocyanate im Wirsing- und Rosenkohl. Collect Czechoslov Chem Commun 24:3606–3610. https://doi.org/10.1135/cccc19593606
PubChem (n.d.) Sulforaphane compound 5350. https://pubchem.ncbi.nlm.nih.gov/compound/5350. Accessed 21 Jan 2025
Ratzka A, Vogel H, Kliebenstein DJ, Mitchell-Olds T, Kroymann J (2002) Disarming the mustard oil bomb. Proc Natl Acad Sci USA 99:11223–11228
Restović I, Kević N, Kurić L, Bočina I, Vuko E, Vrca I (2024) Myrosin cells and myrosinase expression pattern in nasturtium (*Tropaeolum majus* L.). Agronomy 14(9):2108
Robiquet PJ, Boutron F (1831) Nouvelles expériences sur la semence de moutarde. J Pharm Sci Accès 17:279–282
Sangkret S, Pongmalai P, Devahastin S, Chiewchan N (2019) Enhanced production of sulforaphane by exogenous glucoraphanin hydrolysis catalyzed by myrosinase extracted from Chinese flowering cabbage (*Brassica rapa* var. parachinensis). Sci Rep 9(1):9882
Schmid H, Karrer P (1948a) Synthese der racemischen und der optisch aktiven Formen des Sulforaphans. Helv Chim Acta 31(6):1497–1505
Schmid H, Karrer P (1948b) Über Inhaltsstoffe des Rettichs I. Über Sulforaphen, ein Senföl aus Rettichsamen (Raphanus sativus L. var. alba). Helv Chim Acta 31(4):1017–1028
Shirakawa M, Hara-Nishimura I (2018) Specialized vacuoles of myrosin cells: chemical defense strategy in Brassicales plants. Plant Cell Physiol 59(7):1309–1316
Sikorska-Zimny K, Beneduce L (2021) The glucosinolates and their bioactive derivatives in Brassica: a review on classification, biosynthesis and content in plant tissues, fate during and after processing, effect on the human organism and interaction with the gut microbiota. Crit Rev Food Sci Nutr 61(15):2544–2571
Smits JP, Knol W, Bol J (1993) Glucosinolate degradation in liquid and solid-state fermentation. Appl Microbiol Biotechnol 38:696–701
The Human Metabolome Database (n.d.) Metabocard for sulforaphane (HMDB0005792) and sulforaphane-cysteine-glycine (HMDB0240559). https://hmdb.ca/metabolites/. Accessed 28 Jan 2025
Will H, Körner W (1863) Zur Kenntniss der Bildung des Senföls aus dem Samen des schwarzen Senfs. Justus Liebigs Ann Chem 125(3):257–281. https://doi.org/10.1002/jlac.18631250302
Wu H, Liang H, Yuan Q, Wang T, Yan X (2010) Preparation and stability investigation of the inclusion complex of sulforaphane with hydroxypropyl-β-cyclodextrin. Carbohydr Polym 82(3):613–617
Wu Y, Mao J, You Y, Liu S (2014) Study on degradation kinetics of sulforaphane in broccoli extract. Food Chem 155:235–239
Yagishita Y, Fahey JW, Dinkova-Kostova AT, Kensler TW (2019) Broccoli or sulforaphane: is it the source or dose that matters? Molecules 24(19):3593
Yi GE, Robin AHK, Yang K, Park JI, Kang JG, Yang TJ, Nou IS (2015) Identification and expression analysis of glucosinolate biosynthetic genes and estimation of glucosinolate contents in edible organs of *Brassica oleracea* subspecies. Molecules 20(7):13089–13111. https://doi.org/10.3390/molecules200713089
Yin L, Chen C, Chen G, Cao B, Lei J (2015) Molecular cloning, expression pattern and genotypic effects on glucoraphanin biosynthetic related genes in Chinese kale (*Brassica oleracea* var. alboglabra Bailey). Molecules 20(11):20254–20267
Youseif SH, Abdel-Fatah HM, Khalil MS (2022) A new source of bacterial myrosinase isolated from endophytic *Bacillus* sp. NGB-B10, and its relevance in biological control activity. World J Microbiol Biotechnol 38(11):215
Yuanfeng W, Chengzhi L, Ligen Z, Juan S, Xinjie S, Yao Z, Jianwei M (2021) Approaches for enhancing the stability and formation of sulforaphane. Food Chem 345:128771
Zhang Y, Talalay P, Cho CG, Posner GH (1992) A major inducer of anticarcinogenic protective enzymes from broccoli: isolation and elucidation of structure. Proc Natl Acad Sci USA 89(6):2399–2403

Zhang S, Ying DY, Cheng LJ, Bayrak M, Jegasothy H, Sanguansri L, Augustin MA (2020) Sulforaphane in broccoli-based matrices: effects of heat treatment and addition of oil. LWT Food Sci Technol 128:109443

Zhao Z, Chen Q, Qiao X, Wang J, Ali ZTA, Li J, Yin L (2025) Sulforaphane in cancer precision medicine: from biosynthetic origins to multiscale mechanisms and clinical translation. Front Immunol 16:1702860

Zhuang L, Xu K, Zhu Y, Wang F, Xiao J, Guo L (2021) Calcium affects glucoraphanin metabolism in broccoli sprouts under ZnSO4 stress. Food Chem 334:127520

Chapter 2
Pharmacokinetics, Bioavailability, and Molecular Mechanisms of Sulforaphane

Abstract Sulforaphane (SFN), an isothiocyanate generated from glucoraphanin via myrosinase-mediated hydrolysis or gut microbiota metabolism, displays favourable yet highly variable pharmacokinetics, with bioavailability of both SFN and glucoraphanin showing substantial interindividual differences and dietary source influencing time to peak plasma concentration. Absolute bioavailability can reach 82% at low doses, declining at higher doses, possibly due to saturable intestinal transport mechanisms. Vegetable storage, processing, and cooking strongly influence SFN yield through myrosinase inactivation, with direct implications for dietary optimisation and therapeutic translation. Despite rapid systemic clearance, SFN conjugates formed via the mercapturic acid pathway distribute efficiently into target tissues such as prostate, lung, and bladder, with the gastrointestinal tract, liver, kidneys, and blood playing central roles in metabolism and elimination via urinary excretion. Reversible interconversion between SFN and its principal metabolite SFN-NAC in plasma and liver microsomes may function as a prodrug reservoir, prolonging exposure. The R-enantiomer exhibits markedly stronger phase II enzyme-inducing potency than the S-form. SFN disrupts the Keap1–Nrf2 interaction, activating ARE-driven cytoprotective gene transcription and modulating phase I/II detoxification enzymes, while exerting anti-inflammatory effects through inhibition of NF-κB and MAPK signalling. SFN also acts as a concentration-dependent CYP inhibitor, raising drug–drug interaction concerns, and NAMPT has emerged as a novel putative molecular target.

Keywords Sulforaphane bioavailability · Pharmacokinetics · Nrf2/Keap1 pathway · CYP enzyme interactions

2.1 Sulforaphane Pharmacokinetics

Mammals do not synthesise myrosinase; however, the conversion of glucoraphanin can occur in the gastrointestinal tract because of the presence of bacterial microflora (Yagishita et al. 2019; Dmytriv et al. 2025). Within the gastrointestinal tract,

K. Sikorska-Zimny, *Sulforaphane in Human Health*, SpringerBriefs in Food, Health, and Nutrition, https://doi.org/10.1007/978-3-032-34331-4_2

sulforaphane is primarily absorbed in the jejunum, although it has also been detected in the colon of rats. The compound diffuses passively into the bloodstream across cellular membranes, including those of the intestinal epithelium, a process facilitated by the low molecular weight and lipophilic nature of sulforaphane, and is governed by Fick's Law of Diffusion (Shekarri and Dekker 2021; Tortorella et al. 2015; Marshall et al. 2023). Petri et al. (2003) estimated the jejunal absorption of sulforaphane to be approximately 74% ± 29. Once in circulation, sulforaphane binds to the thiol groups of plasma proteins and subsequently enters cells, where it undergoes passive conjugation with glutathione (GSH) catalysed by glutathione *S*-transferases (GSTs) (Kassahun et al. 1997). This rapid conjugation with GSH generates a significant concentration gradient that effectively acts as a driving force for the continued passive diffusion of unconjugated SFN into the tissue (Tortorella et al. 2015). The conjugate is then sequentially metabolised by γ-glutamyl transpeptidase (GGT), cysteinyl-glycinase (GCase), and ultimately *N*-acetyltransferase (NAT) (Petri et al. 2003; Yagishita et al. 2019; Liang et al. 2020; Zuo et al. 2023) (Fig. 2.1).

The reaction between SFN and GSH can occur spontaneously under mild conditions and is significantly accelerated by the enzyme glutathione S-transferase (GST). This enzymatic activity further enhances the rate and extent of cellular accumulation. The rapid conversion of free SFN to its conjugated form allows cells to achieve intracellular concentrations several hundred times higher than extracellular concentrations, often reaching millimolar levels (Zhang and Callaway 2002). Because the resulting GS-SF conjugates are rapidly exported from the cell via membrane transporters such as Pgp-1 (P-glycoprotein/MDR1) and MRP1 (multi-drug resistance-associated protein 1), continuous uptake of free SFN is required to offset this export and maintain high intracellular levels (Zhang and Callaway 2002; Schepici et al. 2020).

Metabolites that circulate systemically, such as sulforaphane-N-acetyl-L-cysteine (SFN-NAC), are excreted in the urine or eliminated in bile conjugated with glutathione (Egner et al. 2011; Bouranis et al. 2024). This metabolic route, known as the mercapturic acid pathway, results in the formation of dithiocarbamate (DTC) derivatives, which are subsequently excreted in the urine (Platz et al. 2015; Zhao et al. 2025).

The initial conjugate is subsequently processed into sulforaphane–cysteinylglycine (SFN-CG), sulforaphane–cysteine (SFN-Cys), and ultimately, sulforaphane–N-acetylcysteine (SFN-NAC) (Platz et al. 2015; Tortorella et al. 2015; Zhao et al. 2025). In human plasma, sulforaphane–glutathione conjugate (SFN-GSH) often accounts for more than 50% of the total sulforaphane metabolites, although free SFN and other derivatives, such as SFN-NAC, SFN-CG, and SFN-Cys, are also present in quantifiable amounts (Saha et al. 2012; Tortorella et al. 2015; Platz et al. 2015; Langston-Cox et al. 2020). In 1998, Shapiro et al. also identified a significant conversion of isothiocyanates to dithiocarbamates within the intestine.

Although urinary excretion is regarded as the primary elimination pathway for sulforaphane, the extent to which it is excreted via faeces remains unclear (Talalay et al. 1995; Tortorella et al. 2015; Yang et al. 2016; Sun et al. 2017; Leone et al. 2017; Liang et al. 2020; Mahn and Castillo 2021; Bouranis et al. 2024).

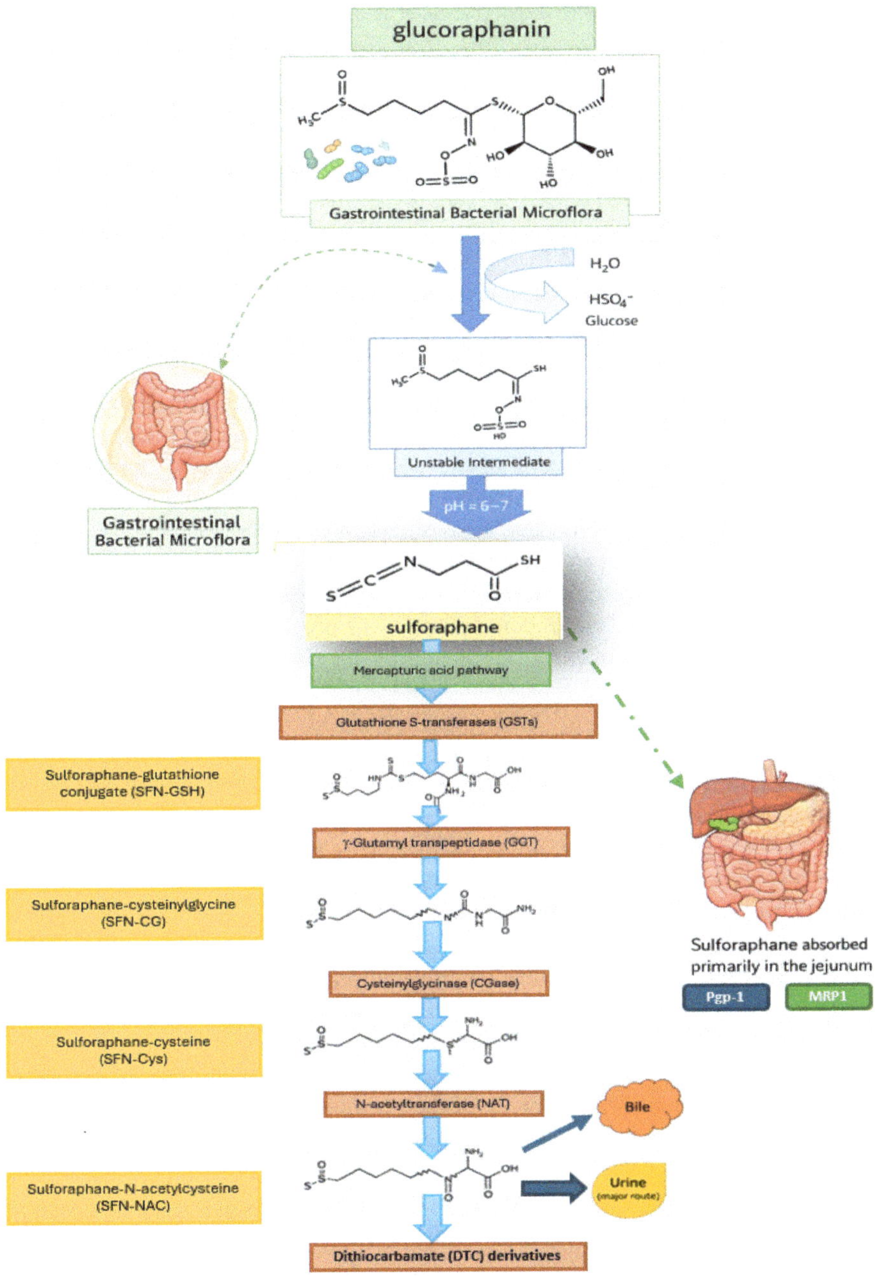

Fig. 2.1 Sulforaphane formulation and metabolism pathways in mammals (Matusheski et al. 2004; Backenköhler et al. 2018; Sikorska-Zimny and Beneduce 2021; HMDB0005792, HMDB0240559 n.d.)

Table 2.1 Domains of Nrf2 (Pańczyszyn-Trzewik et al. 2020; Sajak et al. 2023)

	Role
Neh1	CNC-bZip region enabling the binding of Nrf2 to nuclear DNA
Neh2	Mediates binding of Nrf2 to KEAP1 and targets Nrf2 for ubiquitination and proteasomal degradation
Neh3	Interacts with the chromo-ATPase/helicase DNA-binding protein 6 (CHD6)
Neh4	Transactivation domain that recruits transcriptional co-activators such as CBP/p300
Neh5	Similar but stronger than Neh4 and essential for full transcriptional activation of NRF2 target genes
Neh6	Associated with the regulation of proteasomal degradation
Neh7	Inhibiting the transcriptional activity of Nrf2 and the expression of Nrf2-dependent genes

Sulforaphane can modulate multiple cellular targets simultaneously. It influences carcinogen-metabolising enzymes, thereby affecting the action of mutagens, inhibiting cell proliferation and neoangiogenesis (including metastasis formation), and inducing apoptosis (The Human Metabolome Database n.d.; Fimognari and Hrelia 2007). Detailed investigations have demonstrated that sulforaphane plays a critical role in the suppression of pro-inflammatory mediators through the activation of the Nrf2/ARE (Nuclear factor erythroid 2-related factor 2/antioxidant response element) pathway in an Nrf2-dependent manner. This process occurs via the induction of nuclear translocation of Nrf2, facilitated by the disruption of the Nrf2-Keap1 (Kelch-like ECH-associated protein 1) complex (Tortorella et al. 2015; Townsend and Johnson 2016; Liang et al. 2020; Cazalla et al. 2024).

Nuclear factor erythroid 2-related factor 2 (Nrf2) is a member of the Cap'n'Collar (CNC) subfamily of transcription factors and consists of seven conserved functional domains, designated Neh1–Neh7 (Table 2.1).

Nrf2 is encoded by the NFE2L2 gene, located on the long arm of chromosome 2 (2q31.2), and is predominantly expressed in the kidneys, muscles, and lungs in adults, as well as in the liver and muscles during foetal development (Moi et al. 1994).

Within cells, Nrf2 is typically bound to Kelch-like ECH-associated protein 1 (Keap1). Under conditions of oxidative or electrophilic stress, this complex dissociates, allowing Nrf2 to be actively transported into the nucleus, where it is activated through hetero-dimerisation with small Maf transcription factors (Stępkowski and Kruszewski 2011; Cazalla et al. 2024). The resulting complex binds to the antioxidant response element (ARE/EpRE) in the promoter regions of various phase II detoxification enzymes, thereby modulating cellular defence mechanisms (Tortorella et al. 2015).

Dinkova-Kostova et al. (2002) demonstrated that sulforaphane interacts with thiol groups within Keap1, leading to the disruption of the Keap1-Nrf2 complex and subsequent activation of Nrf2.

The pharmacological activity of sulforaphane is closely linked to its modulatory effects on phase I and II detoxification enzymes (Mahéo et al. 1997; Yagishita et al. 2019; Mangla et al. 2021; Cascajosa-Lira et al. 2024). These enzymatic pathways play crucial roles in the metabolism and excretion of xenobiotics, including drugs and carcinogens.

Table 2.2 Phase I detoxification enzymes (Li et al. 1995; Venkatakrishnan et al. 1998; Ekins and Wrighton 1999; Fretland and Omiecinski 2000; Yu et al. 2001; Edmondson et al. 2004; Strolin Benedetti et al. 2006; Zhou et al. 2009; Go et al. 2015; Kim et al. 2015; Vanduchova et al. 2016; Mokkawes and de Visser 2023)

Enzyme	Family	Primary reaction
CYP1A1/2	CYP 1	Hydroxylation and oxidation of polycyclic aromatic hydrocarbons
CYP1B1		Estrogene hydroxylation
CYP2A6	CYP 2	Specific activity towards nicotine
CYP2B6		Hydroxylation
CYP2C8		Oxidative reactions
CYP2C9		Hydroxylation
CYP2C19		Demethylation
CYP2D6		O- and N-demethylation
CYP2E1		Hydroxylation
CYP3A4	CYP 4	Hydroxylation
FMO1–5	Flavin monooxygenase	Oxygenation
MAO-A/MAO-B	Monoamine oxidase	Oxidative deamination
Epoxide hydrolase (microsomal—mEH and soluble—sEH)	EH	Epoxide hydrolysis
Aldehyde dehydrogenase (ALDH)	ALDH	Aldehyde oxidation
Alcohol dehydrogenase (ADH)	ADH	Alcohol oxidation
Xanthine oxidase (XO)	Molybdenum enzyme	Oxidation of purines

Primary location: Liver (hepatocyte ER), intestine, lung, kidney

Phase I enzymes primarily function by increasing the polarity of hydrophobic compounds through oxidation, reduction, and hydrolysis reactions (Iyanagi 2007). Approximately 75% of all drugs undergo biotransformation via cytochrome P450 (CYP) enzymes (CYP1A1, CYP1A2, CYP1B1, CYP2B6, CYP2C9, CYP2C19, CYP2D6, CYP2E1, and CYP3A4), which facilitate the conversion of pro-carcinogens into their active carcinogenic forms (Table 2.2). The relative balance between the activation and detoxification of these compounds determines their overall carcinogenic potential (Wogan et al. 2004). The remaining phase I enzymes are flavin monooxygenases (FMO), monoamine oxidases (MAO), carbonyl reductases (CRB), molybdenum hydroxylases, and epoxyhydroxylases (Piska and Pękala 2015). Among the key CYP isoenzymes involved in these reactions are CYP2A6, CYP2B6, CYP2C8,9,18, CYP2D6, CYP2E1 and CYP3A4,5 (in human hepatocytes), as well as CYP1A1 and CYP2B1/2 (in rodent models) (Barcelo et al. 1996; Mahéo et al. 1997; Liska 1998; Wogan et al. 2004; Wang et al. 2014; Tortorella et al. 2015). The enzymatic activity of these CYP isoforms promotes the formation of electrophilic intermediates capable of covalently binding to nucleophilic sites within macromolecules, ultimately leading to DNA damage and mutations that initiate carcinogenesis. Notably, sulforaphane inhibits the catalytic activity of these enzymes, thereby exerting chemopreventive effects (Yoxall et al. 2005; Tortorella et al. 2015; Gao et al. 2021).

Phase II enzymes are involved in increasing the polarity of compounds (xenobiotics, drugs) to enable their excretion in urine or faeces (Aronica et al. 2022; Bouranis et al. 2024) (Table 2.3). The upregulation of these enzymes is considered beneficial for enhancing xenobiotic clearance and reducing toxic accumulation. There are nine classes of phase II inducers: 1,2-dithiole-3-thiones; diphenols, phenylenediamines, and quinones; divalent heavy metals; carotenoids, curcumins, and related polyenes; isothiocyanates; hydroperoxides and hydrogen peroxide; Michael reaction acceptor compounds containing olein or acetylene linkages conjugated to electron-withdrawing groups; trivalent arsenicals; and vicinal dimercaptans (Dinkova-Kostova and Talalay 1999; Fahey and Talalay 1999; Khachik et al. 1999).

Phase II enzymes are potent antioxidants that facilitate the excretion of phase I metabolites by binding to them (Tortorella et al. 2015). Isothiocyanates induce phase II enzymes, including haeme oxygenase 1 (HO-1), catalase, glutamate-cysteine ligase (GCL), glutathione-S-transferase (GST), NADPH:quinone dehydrogenase-1 (NQO1), nicotinamide adenine dinucleotide phosphate quinone oxidoreductase, uridine 5′-diphospho (UDP) epoxide hydrolase, superoxide dismutase (SOD), and UDP-glucuronosyltransferase (UGT), through the activation of nuclear factor erythroid 2-related factor 2 (Nrf2) and the antioxidant response element (ARE) (Petri et al. 2003; Iyanagi 2007; Guerrero-Beltrán et al. 2012; de Figueiredo et al. 2015; Su et al. 2018; Bouranis et al. 2024). Sulforaphane has been shown to induce glutathione transferases, NAD(P)H:quinone reductase (DT-diaphorase), and haeme oxygenase (Talalay et al. 1995; Mahéo et al. 1997; Tortorella et al. 2015; Wang et al. 2014; Mangla et al. 2021; Thiruvengadam et al. 2021; Feng et al. 2023).

Analysis of human plasma and urine from volunteers consuming cruciferous vegetables revealed increased levels of Nrf2, GSTs, and NQO1 (Yagishita et al. 2019).

The anti-inflammatory properties of sulforaphane involve multiple complex signalling pathways, with the nuclear factor kappa-light-chain-enhancer of activated B-cells (NF-κB) and Mitogen-Activated Protein Kinase (MAPK) pathways being among the most significant (Keum et al. 2006; Múnera-Rodríguez et al. 2024). Under physiological conditions, the NF-κB complex remains inactive in the cytoplasm because of its association with the inhibitor of nuclear factor kappa B (IκB), which prevents its translocation to the nucleus. Upon exposure to xenobiotics or pathogens, this complex becomes inactivated, leading to the activation of IκB kinase (IKK), phosphorylation of IκB, and its subsequent degradation. This process ultimately results in the release and nuclear translocation of NF-κB, where it exerts transcriptional activity on its target genes. The NF-κB pathway regulates the production of pro-inflammatory cytokines and mediates the expression of key inflammatory molecules, such as nitric oxide (NO), inducible nitric oxide synthase (iNOS), cyclooxygenase-2 (COX-2), and prostaglandin E2 (PGE2) (Qi et al. 2016; Subedi et al. 2019; Mangla et al. 2021; Múnera-Rodríguez et al. 2024). Sulforaphane binds to the NF-κB complex.

Table 2.3 Phase II detoxification enzymes (Beall et al. 1995; Radominska-Pandya et al. 1999; Hein et al. 2000; Cannon et al. 2002; Siegel et al. 2004; Bai et al. 2007; Dourado et al. 2010; Cook 2011; Gan et al. 2024)

Enzyme	Abbreviation	Conjugation reaction	Group donor
Glutathione S-transferase	GST (cytosolic: GSTA, GSTM, GSTP, GSTT; mitochondrial: GSTK; microsomal: MGST/MAPEG)	Nucleophilic addition of glutathione (designated cytosolic, microsomal, and mitochondrial transferases)	GSH
UDP-glucuronosyltransferase	UGT	Glucuronidation	UDP-glucuronic acid
Sulfotransferase	SULT1A1, SULT1E1, SULT2A1	Sulfation (SULT1A1—Exogenous substances; SULT1E1, SULT2A1—Endogenous substances)	PAPS (3′-phosphoadenosine-5′-phosphosulfate)
N-acetyltransferase	NAT1/NAT2	N-acetylation	Acetyl-CoA
Thiopurine methyltransferase	TPMT	Aromatic thiol methylation	SAM (S-adenosylmethionine)
Catechol-O-methyltransferase	COMT	O-methylation of catechols	SAM or AdoMet
Amino acid conjugase	–	Glycine/taurine conjugation	Glycine, taurine
Glutamate-cysteine ligase (Phase II response enzymes)	GCL (GCLc + GCLm)	GSH synthesis (rate-limiting)	Glutamate + cysteine + ATP
NAD(P)H:quinone oxidoreductase (Phase II response enzymes)	NQO1/DTD/DT-diaphorase	Two-electron reduction of quinones	NADPH/NADH
Heme oxygenase (Phase II response enzymes)	HO-1/HO-2	Heme degradation → biliverdin	Heme + NADPH + O_2

Primary location: Liver (cytosol + ER), intestine, kidney, lung

The MAPK signalling pathway comprises a family of protein kinases that mediate cellular responses by transducing inflammatory signals, ultimately inducing the production of pro-inflammatory cytokines, such as interleukin-6 (IL-6), interleukin-8 (IL-8), and tumour necrosis factor-α (TNF-α) (Moens et al. 2013; Sabio and Davis 2014). Sulforaphane modulates this pathway by influencing extracellular signal-regulated kinases (ERK), p38 MAP kinases, and c-Jun NH2-terminal kinases (JNK) (Peng et al. 2015; Subedi et al. 2019; Xie et al. 2021; Múnera-Rodríguez et al. 2024).

Through this mechanism, sulforaphane inhibits the pro-inflammatory cytokines IL-1β and IL-6 in primary microglia from aged mice, as well as TNF-α production (Townsend and Johnson 2016; Mohanty et al. 2018). The primary role of IL-1β in the pro-inflammatory response is to facilitate communication between the immune system and brain. Additionally, it influences IL-6 production. Both cytokines contribute to increased body temperature and can cross the blood-brain barrier (BBB), where their receptors are located (Karpel 2001; Czerwiec et al. 2016) (Fig. 2.2).

2.2 Bioavailability

Research has highlighted significant differences in the metabolism of glucosinolates (glucoraphanin) and isothiocyanates (ITCs) in the human body, demonstrating that sulforaphane (SFN) exhibits significantly higher bioavailability—exceeding 70%—compared to its precursor, glucoraphanin (Montgomery and Srinivasan 2019). Egner et al. (2011) estimated the bioavailability of sulforaphane to be 70%, whereas that of glucoraphanin was only 5%, and noted lower interindividual variability in sulforaphane absorption. The superior bioavailability of SFN compared to that of GRA is attributed to the enzymatic degradation of glucosinolates by the human gut microbiota (Li et al. 2011; Luang-In et al. 2014; Bouranis et al. 2021; Montgomery and Srinivasan 2019). However, as pointed out by Li et al. (2011), "*no direct link to specific bacterial species could be established, possibly due to the complexity and functional redundancy of the gut microbiota.*" Therefore, precision medicine approaches emphasise host factors, such as the role of the gut microbiome in converting glucoraphanin to active sulforaphane and host genetic polymorphisms in glutathione S-transferase (GST) genes. Individuals with GSTM1-null or GSTT1-null genotypes exhibit higher and more prolonged levels of bioactive sulforaphane in the bloodstream, identifying them as "*optimal responders*" to chemopreventive interventions (Zhao et al. 2025).

The time required for sulforaphane to reach peak concentrations in biological fluids varies depending on its source and formulation. Reported values include 1 h for plasma following ingestion of broccoli sprout powder (Grady et al. 2023), 1.5–2 h for plasma after broccoli or super broccoli consumption (Gasper et al. 2005), 2 h for plasma/urine following intake of broccoli extract supplements or broccoli sprouts (Langston-Cox et al. 2020; van Steenwijk et al. 2023), and 3–6 h for urinary excretion after broccoli sprout consumption (Atwell et al. 2015). In a study investigating the effects of a single serving of fresh broccoli microgreens,

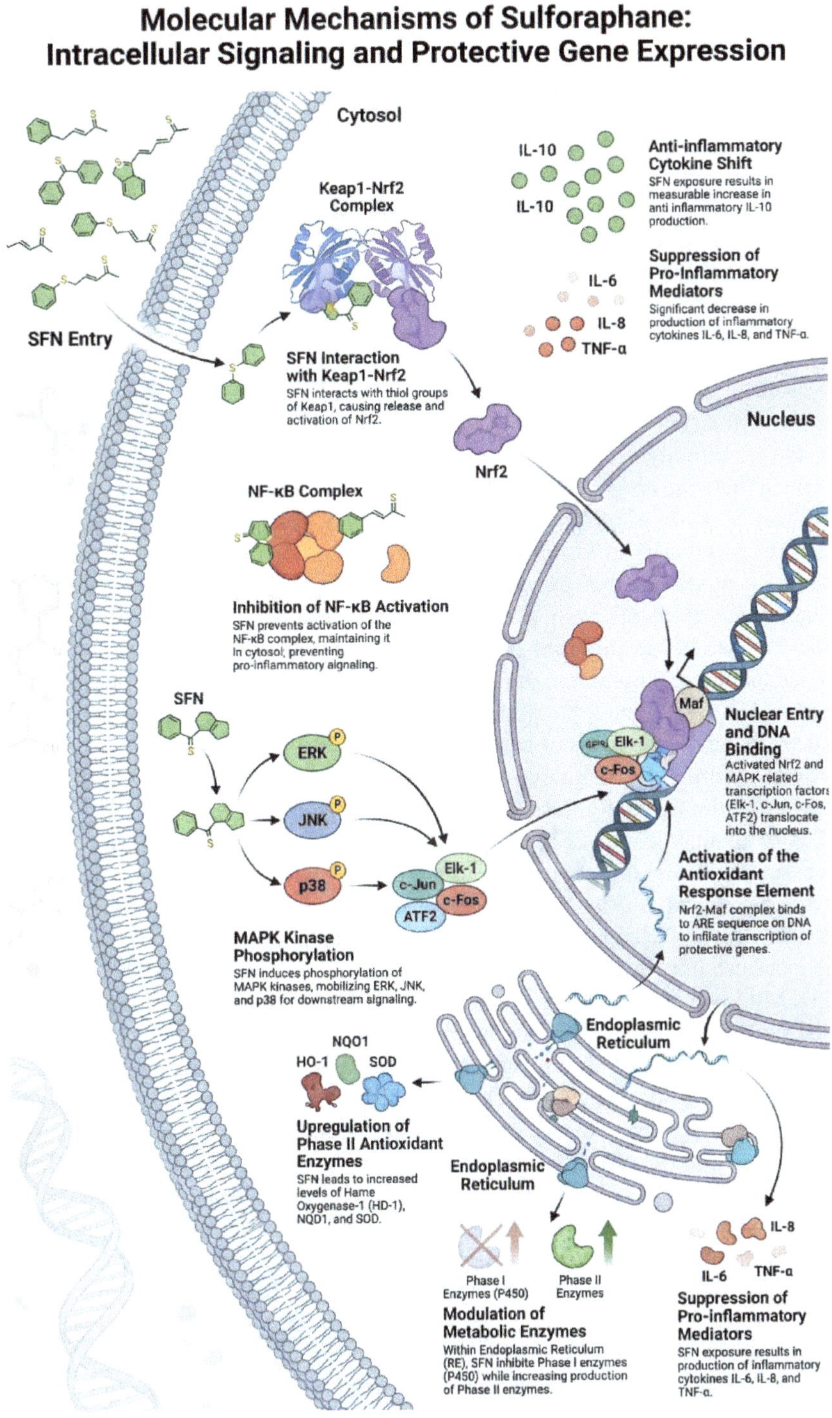

Fig. 2.2 Mechanism of action of sulforaphane in cells (This image has been enhanced using artificial intelligence (AI) tools (Google, NotebookLM, 2026)

SFN metabolite excretion in urine peaked at 3–6 h post-consumption, with SFN-N-acetyl-L-cysteine being the most abundant metabolite detected in the early stages (Bouranis et al. 2023). Despite their rapid clearance from the blood, these conjugates are effectively distributed to target organs (e.g. the prostate, lung, and bladder), where they can exert biological and therapeutic effects (Shekarri and Dekker 2021; Livingstone 2022; Zhao et al. 2025) (Fig. 2.3). Interestingly, sulforaphane-nitrile (SFN-NIT) excretion followed a different kinetic profile, peaking between 6 and 24 h and remaining detectable even 48 h after ingestion. While urine remains the primary route of excretion, recent research has successfully detected SFN metabolites in human stool for the first time, although this represents a minor excretion pathway with significantly lower metabolite abundance compared to urine (Bouranis et al. 2023).

Nevertheless, researchers such as Razis et al. (2011) and Giacoppo et al. (2015) suggested focusing specifically on the R-sulforaphane enantiomer (R-SNF) rather than the commonly studied racemic mixture (RS-SNF). Giacoppo et al. (2015) explained that "*the configuration of the sulfoxide stereogenic centre in the GRA side chain was recently ascertained by NMR to be RS, a configuration retained in the hydrolysis product R-sulforaphane.*" Razis et al. (2011) further noted that R-SNF exhibits a much greater capacity to stimulate detoxifying enzymes than the S-enantiomer. Both research teams emphasised the necessity of conducting new studies on R-SNF, as previous investigations that included both enantiomers may have led to underestimation or unreliable conclusions regarding its biological effects (Razis et al. 2011; Giacoppo et al. 2015). Licznerska et al. (2021) pointed out differences in the mechanisms of action of R and S sulforaphane. Vergara et al. (2008) discovered with NMR methods, in *Arabidopis thaliana* a pure epimer with a sulfoxide group of the R_S configuration.

The bioavailability of SFN is profoundly influenced by the storage conditions, processing, and cooking of cruciferous vegetables, which can denature the essential enzyme myrosinase. Raw broccoli consumption has been observed to provide approximately 37% bioavailability, whereas cooked broccoli, where myrosinase has been inactivated by heat, yields only approximately 3.4% bioavailability (Men et al. 2023). Mild heating led to the greatest ITC metabolite concentrations in vivo, followed by the non-heated and steamed broccoli sprout powder diets, with erucin appearing to be the favoured form in the liver, kidney, and bladder, even when only sulforaphane was consumed (Bricker et al. 2014). In addition, storage conditions influence SFN content. Storage of Brussels sprouts at −1.5 °C was more beneficial for the preservation of SFN than storage at 5 °C (Sikorska-Zimny and Badełek 2014). Bouranis et al. (2023) recommended the consumption of raw, unprocessed broccoli microgreens or sprouts to maximise the delivery of bioavailable SFN. Furthermore, pharmaceutical and supplemental formulations have shown variable success; for instance, SFN-rich extracts have demonstrated higher bioavailability and lower interindividual variability than GRA-rich extracts (Men et al. 2023). In clinical settings, myrosinase-activated broccoli seed extract formulations have been shown to afford significantly better bioavailability than non-activated preparations, with maximum peak levels at 2 h (Langston-Cox et al. 2020).

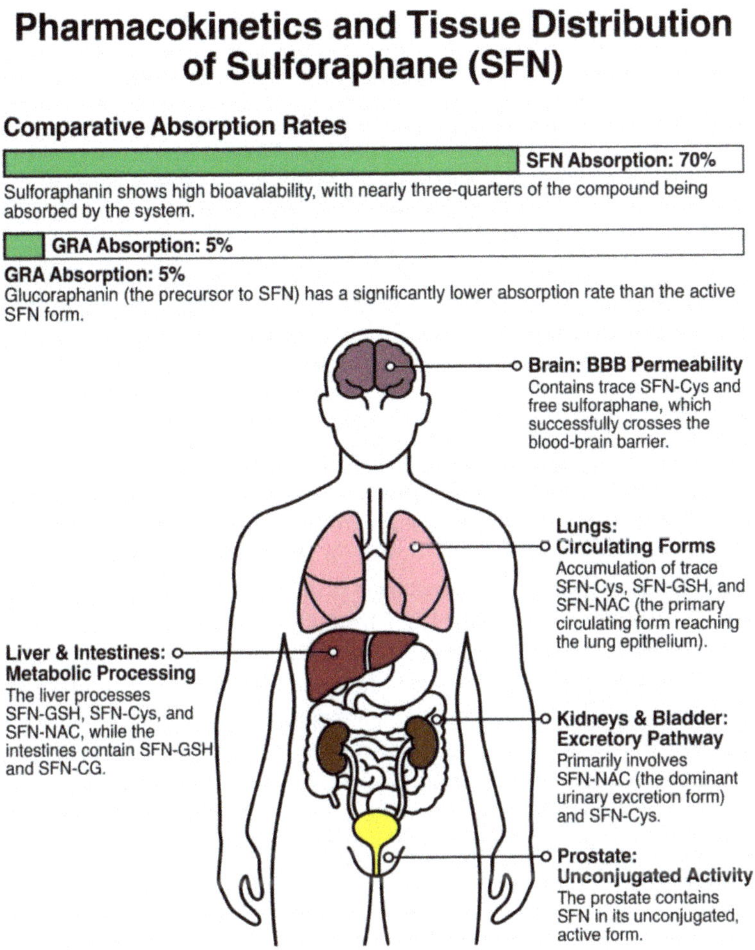

Excretion Profile

	Timeframe	Description
Unchanged SFN (70-90%)	2–6h	The vast majority of SFN is excreted in its original form within the first few hours post-ingestion.
SFN Metabolites (1-45%)	< 6h	Primarily consists of SFN-NAC, representing the processed fraction of the compound.
SFN-NIT	6–24(48)h	A delayed excretion phase for the SFN-NIT metabolite, extending up to two full days.

Fig. 2.3 GRA and SFN bioavailability (Kassahun et al. 1997; Petri et al. 2003; Clarke et al. 2011; Hauder et al. 2011; Bricker et al. 2014; Parfenova et al. 2020; Son et al. 2021; Livingstone 2022; Yan and Yan 2023; Ross 2024), (this image has been enhanced using artificial intelligence (AI) tools (Google, NotebookLM, 2026)

Pharmacokinetic studies have revealed that SFN exhibits nonlinear kinetics (Son et al. 2021). While SFN is well absorbed in the intestine, reaching up to 82% absolute bioavailability at low doses, this percentage decreases as the dose increases, likely due to the saturation of the absorption process (Son et al. 2021). Interestingly, the metabolic relationship between SFN and its major metabolite, SFN-NAC, is reversible in humans. SFN-NAC can be converted back to its parent form, SFN, in both the plasma and liver microsomes, effectively acting as a prodrug. Despite SFN-NAC having lower intestinal absorption and lower oral bioavailability (approximately 24.8–65% for the combination of GRP-rich powder with fresh broccoli sprouts) than SFN, this interconversion allows for the potential development of SFN-NAC formulations to achieve prolonged pharmacological effects while minimising the gastrointestinal side effects associated with high doses of SFN (Cramer et al. 2012; Son et al. 2021).

Biological status and physiological conditions also play critical roles in SFN distribution and absorption (Langston-Cox et al. 2020). Clinical evidence indicates that the therapeutic impact of sulforaphane is not uniform but varies significantly depending on the specific pathophysiology of the targeted disease (Mazarakis et al. 2020). While SFN generally operates through the Nrf2-Keap1 pathway to bolster antioxidant defence, its clinical success and primary mode of action depend on the condition (Elbarbry and Elrody 2011; Conzatti et al. 2015). For example, comparative studies have shown that pregnant women, particularly those with pregnancy hypertension or preeclampsia, exhibit significantly lower plasma levels and lower SFN metabolites than non-pregnant women receiving an identical dose. It has been observed that to achieve therapeutic plasma levels in women with preeclampsia comparable to those in non-pregnant individuals, a doubling of the dose is required (Langston-Cox et al. 2020). Additionally, dosing frequency is a vital factor for maintaining efficacious concentrations; pharmacokinetic evidence suggests that twice-daily dosing of SFN-rich sprout extract retains greater plasma metabolite levels than a single 24-h dose (ElKhalifa et al. 2023). In Type 2 diabetes, SFN demonstrates its most consistent clinical efficacy by directly addressing metabolic dysregulation. Interventions using broccoli sprout powder have led to a significant decrease in fasting blood glucose (averaging 30 mg/dl), reduced hepatic glucose, and improved insulin resistance, particularly in obese patients (Conzatti et al. 2015; Axelsson et al. 2017; Mazarakis et al. 2020). In contrast, the cardiovascular effects of SFN are highly dependent on baseline health status. While SFN significantly improves lipid profiles (total cholesterol, low-density lipoprotein, and triglycerides) and reduces oxidised LDL levels in diabetic cohorts, identical doses often fail to produce changes in blood pressure or endothelial function in hypertensive patients or those at moderate cardiovascular risk (Christiansen et al. 2010; Bahadoran et al. 2012; Armah et al. 2013). The efficacy of SFN in respiratory diseases is characterised by high interindividual variability and genetic dependency.

In asthmatic populations, SFN results in a "*mixed response*" with approximately 60% of patients experiencing a significant bronchoprotective effect, while 20% may experience a worsened response (Mazarakis et al. 2020). This divergence may be linked to the differential induction of the NQO1 enzyme and the underlying GSTM1 genotype, suggesting that SFN's ability to scavenge reactive oxygen species (ROS) from air pollutants is contingent on a patient's genetic capacity to upregulate Phase II enzymes (Riedl et al. 2009; Brown et al. 2015). In patients with chronic obstructive pulmonary disease (COPD), SFN may not significantly alter systemic inflammatory markers but exhibits a unique therapeutic role by improving the phagocytic clearance of bacteria (e.g. *Haemophilus influenzae*) in lung macrophages, a function not typically observed in other cohorts (Harvey et al. 2011; Mazarakis et al. 2020). In gastrointestinal disorders, specifically *Helicobacter pylori* infection, SFN's action shifts from a molecular signalling modulator to a direct antibacterial agent. Unlike its antioxidant role in the lungs, SFN acts by inactivating urease, which is vital for *H. pylori* survival in the acidic gastric environment (Fahey et al. 2013). Clinical studies have shown that daily consumption of SFN-rich sprouts can reduce bacterial colonisation and attenuate gastritis, although these effects may return to baseline after cessation of consumption (Yanaka et al. 2009; Mazarakis et al. 2020). Finally, in oncology, SFN's role is primarily chemopreventive and site-specific. For example, it has not yet shown the ability to drastically reduce prostate-specific antigen (PSA) levels by more than 50% in prostate cancer; it significantly prolongs PSA doubling time (from 6.1 to 9.6 months), indicating a slowing of disease progression rather than a cure (Alumkal et al. 2015). In breast cancer tissue, the action is even more selective; SFN inhibits cellular proliferation in ductal carcinoma in situ (DCIS) but has no significant effect on benign or invasive ductal carcinoma tissue (Zhang et al. 2016). This underscores the conclusion that SFN's clinical utility of SFN is highly specific to the disease stage and unique molecular environment of the affected tissue (Elbarbry and Elrody 2011; Mazarakis et al. 2020).

The source of SFN is a major determinant of its initial content. Broccoli sprouts (3–5-day-old plants) and seeds are the primary sources of SFN, with SFN content in sprouts reaching up to 396.18 mg/100 g FW (Men et al. 2023). Recent evidence suggests that 10-day-old broccoli microgreens have a GRA content (approximately 13 μmol/g) similar to that of broccoli sprouts and remain stable during cold storage, confirming them as a significant and reliable source of bioavailable SFN (Bouranis et al. 2023). To optimise the health benefits derived from these sources, the pharmaceutical industry is exploring enteric-coated formulations and drug delivery systems, such as liposomes and nanoparticles, to overcome SFN's inherent instability and rapid degradation (ElKhalifa et al. 2023; Men et al. 2023).

Research on the accumulation of SNF metabolites has demonstrated that the highest concentrations of this compound are found in tissues where SNF exhibits a strong ability to bind glutathione and efficiently undergoes passive diffusion (Zuo et al. 2023). These findings indicate that the gastrointestinal tract, liver, kidneys, and blood are key sites of SNF metabolism (Clarke et al. 2011). SNF is rapidly absorbed, with a half-life of only a few hours, and is predominantly excreted in the urine, accounting for 70–90% of the administered dose (Yagishita et al. 2019). This characterises SNF as a well-absorbed compound; however, owing to its high instability, ongoing research aims to enhance its stability or administer it in the form of its precursor, glucoraphanin (Yagishita et al. 2019). Nonetheless, the urinary excretion of SNF metabolites has been reported to vary widely, with extreme cases ranging from 1 to 45% (Yagishita et al. 2019) (Fig. 2.3).

One of the most effective methods for quantifying sulforaphane (in the picomolar range) and its metabolites (dithiocarbamates) is cyclocondensation with 1,2-benzenedithiol and liquid chromatography coupled with tandem mass spectrometry (LC-MS/MS), either alone or in combination with stable isotope-labelled internal standards of sulforaphane (Yagishita et al. 2019).

2.3 Effects with Drugs

Several studies have highlighted the synergistic effects of sulforaphane in combination with cytostatic drugs, such as gemcitabine and cisplatin, leading to enhanced cancer cell death. This effect is likely attributed to the prolonged intracellular retention of these drugs in malignant cells (Sharma et al. 2011; Calcabrini et al. 2020; Gong et al. 2020; Khatoon et al. 2022). Notably, the co-administration of doxorubicin and SFN inhibits triple-negative breast cancer growth while simultaneously reducing cardiotoxicity (Khatoon et al. 2022).

In lung cancer, the combination of gefitinib and SFN has been reported to overcome gefitinib resistance by inhibiting the sonic hedgehog signalling pathway (Wang et al. 2018, 2021). Similarly, SFN combined with cisplatin in advanced epidermal squamous cell carcinoma has been found to delay tumour progression (Kerr et al. 2017).

The anti-cancer potential of SFN is further enhanced when used in combination with natural compounds. For instance, SFN and epigallocatechin gallate have been suggested to resensitise oestrogen (ER)-negative breast cancer cells to tamoxifen (Li et al. 2017). Additionally, SFN combined with curcumin and aspirin has demonstrated chemosensitisation effects in pancreatic cancer cells, inhibiting proliferation and inducing apoptosis (Sutaria et al. 2012). Furthermore, in pancreatic cancer cell lines, the anti-cancer efficacy of loratadine was significantly enhanced by SFN co-treatment (Desai et al. 2019).

The literature does not report any significant adverse effects or interactions between sulforaphane and other drugs (Zhang et al. 2014). However, DrugBank has listed two potential consequences of drug interactions—methemoglobinemia and thrombosis—based on information extracted from drug labels and scientific publications (DrugBank n.d.). The risk or severity of methemoglobinemia can be increased when sulforaphane is combined with ambroxol, articaine, benzocaine, benzyl alcohol, bupivacaine, butacaine, butamben, capsaicin, chloroprocaine, cinchocaine, cocaine, diphenhydramine, dyclonine, ethyl chloride, etidocaine, levobupivacaine, lidocaine, meloxicam, mepivacaine, oxetacaine, oxybuprocaine, phenol, pramocaine, prilocaine, procaine, proparacaine, propoxycaine, ropivacaine, and tetracaine. The risk or severity of thrombosis can be increased when darbepoetin alfa, erythropoietin, methoxy polyethylene glycol-epoetin beta, or peginesatide is combined with sulforaphane.

A critical area of research is the transition of SFN from a dietary supplement to a potential clinical therapeutic agent, with the ability to interact with conventional pharmaceutical drugs. These interactions can be synergistic, enhancing therapeutic outcomes and reducing toxicity, or antagonistic, potentially leading to drug resistance or altered pharmacokinetics (Lubelska et al. 2012).

One of the most promising areas of drug-SFN interaction research is the management of chronic pain syndromes, particularly fibromyalgia (FM). Current pharmacological treatments for FM, such as gabapentinoids (gabapentin and pregabalin), are often limited by significant adverse effects, including drowsiness, dizziness, and ataxia. Studies using murine models have demonstrated that the combination of SFN and gabapentin (GBP) produces a significant synergistic antiallodynic and antihyperalgesic effect (Zamora-Díaz et al. 2025). Specifically, experimental data suggest that combining intermediate doses of SFN and GBP allows for a reduction in the required dosage of each compound to approximately one-third of their individual effective doses while maintaining the same level of pain relief. Molecular docking and electrocorticographic (ECoG) analyses indicated that this synergy may be mediated by the interaction of both compounds with the $\alpha 2\delta$ subunit of voltage-gated calcium channels. SFN appears to function as a regulator of excitatory synaptic transmission, potentially enhancing the inhibitory effects of gabapentin in the central nervous system (Zamora-Díaz et al. 2025).

In the context of Alzheimer's disease (AD), SFN has been studied for its ability to act in conjunction with nonsteroidal anti-inflammatory drugs (NSAIDs). Bioinformatics and in silico studies suggest that dexibuprofen, the active enantiomer of ibuprofen, may exhibit synergistic effects when co-administered with SFN. This combination targets neuroinflammation, insulin resistance, and apoptotic signalling pathways, which are pivotal in the pathogenesis of AD (Vu and Nguyen 2024). SFN further complements this by reducing amyloid-β accumulation and mitigating oxidative stress via the Nrf2 pathway (Mangla et al. 2021).

SFN is widely recognised for its ability to potentiate the efficacy of various chemotherapeutic drugs while simultaneously offering protective effects to healthy tissues. This dual-action profile makes it an ideal candidate for combination therapy in oncology (Kamal et al. 2020). An example of such activity can be drawn with taxanes (paclitaxel and docetaxel), particularly in treating triple-negative breast cancer (TNBC), which often induces the expansion of cancer stem cells (CSCs) by triggering the secretion of interleukin-6 (IL-6) (Yuan et al. 2018). However, SFN effectively eliminates CSCs by inhibiting the NF-κB survival pathway (Kamal et al. 2020; Mangla et al. 2021). Similarly, SFN enhances gemcitabine potency in breast cancer models. Studies have reported a 40-fold reduction in the IC_{50} of loratadine when combined with SFN-enabled delivery systems, significantly increasing the induction of apoptosis (Kamal et al. 2020). The combination of SFN and 5-flurouracil (5-FU) also demonstrated high efficacy. In models of salivary gland adenoid cystic carcinoma and colon cancer, SFN significantly enhanced the inhibitory effects of 5-FU on cell proliferation. This synergism is partly attributed to SFN's ability of SFN to induce cell cycle arrest in the G2/M phase (Kamal et al. 2020).

Cisplatin resistance is a major clinical hurdle in the treatment of ovarian and squamous cell carcinomas. SFN restores cisplatin sensitivity by regulating DNA repair mechanisms and increasing intracellular drug accumulation (Mangla et al. 2021). In squamous cell carcinoma cells (SCC-13 and HaCaT), co-administration of SFN and cisplatin inhibited cell proliferation and migration and the formation of stem cell spheroids. Furthermore, SFN sensitises HER2-overexpressing breast cancer cells to lapatinib and reverses resistance to gefitinib in lung cancer models by modulating the Sonic hedgehog signalling pathway (Kamal et al. 2020).

While many SFN-drug interactions are therapeutic, SFN can significantly influence the pharmacokinetics of co-administered drugs by modulating cytochrome P450 (CYP) enzymes and membrane transporters. This modulation can lead to unexpected drug-drug interactions (DDIs). Systematic analyses of human hepatocytes and liver microsomes have revealed that SFN is a concentration-dependent inhibitor of several key CYP enzymes in humans. Specifically, SFN inhibits CYP2A6, CYP2D6, and CYP3A4 (Srovnalova et al. 2015). Inhibition of CYP3A4 is particularly significant, as this enzyme is responsible for the metabolism of 30–50% of clinically used drugs (Zhang et al. 2024). Research has shown that SFN acts as an antagonist of the Pregnane X Receptor (PXR), thereby preventing the induction of CYP3A4 by other drugs, such as rifampicin (Srovnalova et al. 2015).

The interaction between SFN and commonly consumed drugs, such as verapamil, furosemide, and ketoprofen, can alter their transport and elimination. Studies on Caco-2 cell models have shown that SFN interacts antagonistically with these drugs by modulating Phase II and III enzymes (Lubelska et al. 2012). For instance, SFN can increase the expression of multi-drug resistance protein 1 (MRP1), a Phase III transporter (Lubelska et al. 2012). The metabolites that "passed" Phases I and II are then handled by Phase III, which comprises membrane-bound transport proteins responsible for their active export from the cell. Phase III transporters include multi-drug resistance-associated protein (MRP) and P-glycoprotein (PgP), both of which belong to the ATP-binding cassette (ABC) protein superfamily (Lubelska et al. 2012). These transporters use ATP hydrolysis to drive the efflux of conjugated metabolites and drugs across cellular membranes, playing a key role in determining the final disposition, bioavailability, and systemic elimination of xenobiotics from the body. The clinical relevance of this transport system is particularly evident in the context of drug resistance, where the upregulation of MRP and PgP mediates the efflux of chemotherapeutic agents before they can exert their cytotoxic effects. While verapamil is often used to block MRP1 to overcome such resistance, co-administration with SFN may paradoxically increase the efficiency of these membrane transporters, thereby abolishing the verapamil effect and potentially contributing to the development of multi-drug resistance, an important consideration when evaluating sulforaphane as an adjunct to conventional chemotherapy. Furthermore, while furosemide and verapamil have been shown to induce the translocation of Nrf2 to the nucleus, SFN, despite being a potent Nrf2 inducer, can reduce the intensity of this translocation when combined with these drugs, suggesting an interaction at the initial level of enzyme regulation (Lubelska et al. 2012).

A significant beneficial interaction of SFN is its ability to protect vital organs from the toxic side effects of intensive drug therapy. Doxorubicin (DOX) is a highly effective anti-cancer agent but is limited by dose-dependent cardiotoxicity caused by oxidative stress. SFN protects cardiomyocytes by restoring the cardiac expression of Nrf2-regulated antioxidant genes and improving mitochondrial respiratory complex activity (Kamal et al. 2020; Mangla et al. 2021). SFN has the potential to reduce the hepatotoxicity and cardiotoxicity associated with tamoxifen treatment in patients with breast cancer. By inhibiting Histone Deacetylases (HDACs) and enhancing Nrf2 activity, SFN improves the safety profile of tamoxifen while potentially potentiating its anti-tumour activity (Mangla et al. 2021).

SFN's anti-obesity mechanisms are significantly modulated by the antimalarial drug chloroquine (also used as an immunomodulatory drug for rheumatoid diseases). In high-fat diet models, chloroquine intensifies the effects of low-dose SFN (Ashmawy et al. 2022). This interaction is mediated by the modulation of the JAK-2/STAT-3/SOCS-3 inflammatory pathway and autophagy machinery. Chloroquine enhances SFN's ability of SFN to reduce weight gain, lower blood glucose, and improve lipid profiles by inhibiting lipogenesis-related enzymes and downregulating pro-inflammatory markers, such as NF-κB and IL-22 (Ashmawy et al. 2022).

Additionally, bioinformatics research has identified nicotinamide phosphoribosyltransferase (NAMPT) as a potential novel target of SFN (Lagunas-Rangel 2024). SFN is predicted to function as an allosteric activator of NAMPT, mimicking the behaviour of known NAMPT activators. Since NAMPT is a rate-limiting enzyme in NAD^+ metabolism (Zhang et al. 2025), this interaction could have profound implications for the treatment of metabolic disorders and ageing-related diseases (Lagunas-Rangel 2024).

The therapeutic potential of sulforaphane is further amplified when it is used in combination with other bioactive nutraceuticals. Synergistic effects have been observed when sulforaphane is co-administered with curcumin or vitamin D, leading to enhanced HO-1 expression and more robust inhibition of inflammatory markers, such as iNOS and COX-2 (Ruhee and Suzuki 2024).

The interactions between sulforaphane and pharmaceutical drugs are complex and vary significantly, depending on the therapeutic context. In pain management and oncology, SFN often acts as a potent synergist, allowing for reduced drug dosage and decreased toxicity (Kamal et al. 2020; Zamora-Díaz et al. 2025). Its role in activating the Nrf2 pathway provides a robust defence against the oxidative damage induced by drugs (like doxorubicin) (Mangla et al. 2021). However, its ability to inhibit key CYP450 enzymes and modulate MRP1 transporters necessitates caution, as these effects can lead to altered drug clearance and potentially promote multi-drug resistance (Lubelska et al. 2012; Srovnalova et al. 2015).

Despite the abundance of preclinical data, there is a clear need for high-quality clinical trials to confirm the safety and efficacy of SFN-drug combinations in humans (Mangla et al. 2021). Furthermore, the development of nanoengineered drug delivery systems is essential to overcome the inherent biopharmaceutical limitations of SFN, such as poor aqueous solubility and stability, to ensure its full therapeutic potential in clinical settings (Kamal et al. 2020).

References

Alumkal JJ, Slottke R, Schwartzman J, Cherala G, Munar M, Graff JN et al (2015) Phase II study of sulforaphane-rich broccoli sprout extract in recurrent prostate cancer. Investig New Drugs 33:480–489

Armah CN, Traka MH, Dainty JR, Defernez M, Janssens A, Wing L et al (2013) High-glucoraphanin broccoli modulates mitochondrial function. Am J Clin Nutr 98:712–722

Aronica L, Ordovas JM, Volkov A, Lamb JJ, Stone PM, Minich D et al (2022) Genetic biomarkers of metabolic detoxification. Nutrients 14:768

Ashmawy AI, El-Abhar HS, Abdallah DM, Ali MA (2022) Chloroquine modulates sulforaphane anti-obesity mechanisms. Eur J Pharmacol 927:175066

Atwell LL, Hsu A, Wong CP, Stevens JF, Bella D, Yu TW et al (2015) Absorption and chemopreventive targets of sulforaphane in humans. Mol Nutr Food Res 59:424–433

Axelsson AS, Tubbs E, Mecham B, Chacko S, Nenonen HA, Tang Y et al (2017) Sulforaphane reduces hepatic glucose production in type 2 diabetes. Sci Transl Med 9:4477

Backenköhler A, Eisenschmidt D, Schneegans N, Strieker M, Brandt W, Wittstock U (2018) Iron as cofactor of specifier proteins in glucosinolate breakdown. PLoS One 13:e0205755

Bahadoran Z, Mirmiran P, Hosseinpanah F, Rajab A, Asghari G, Azizi F (2012) Broccoli sprout powder improves triglycerides and oxidized LDL. Diabetes Res Clin Pract 96:348–354

Bai HW, Shim JY, Yu J, Zhu BT (2007) O-methylation of catechol substrates by COMT. Chem Res Toxicol 20:1409–1425

Barcelo S, Gardiner JM, Gescher A, Chipman JK (1996) CYP2E1-mediated anti-genotoxicity of sulforaphane. Carcinogenesis 17:277–282

Beall HD, Murphy AM, Siegel D, Hargreaves RH, Butler J, Ross D (1995) NQO1 as target for bioreductive antitumor quinones. Mol Pharmacol 48:499–504

Bouranis JA, Beaver LM, Ho E (2021) Metabolic fate of dietary glucosinolates: role of microbiome. Front Nutr 8:748433

Bouranis JA, Wong CP, Beaver LM, Uesugi SL, Papenhausen EM, Choi J et al (2023) Sulforaphane bioavailability after broccoli microgreens consumption. Foods 12:3784

Bouranis JA, Beaver LM, Wong CP, Choi J, Hamer S, Davis EW et al (2024) Sulforaphane and sulforaphane-nitrile metabolism in humans: microbiome associations. Mol Nutr Food Res 68:2300286

Bricker GV, Riedl KM, Ralston RA, Tober KL, Oberyszyn TM, Schwartz SJ (2014) Isothiocyanate metabolism, distribution, and interconversion in mice following consumption of thermally processed broccoli sprouts or purified sulforaphane. Mol Nutr Food Res 58(10):1991–2000

Brown RH, Reynolds C, Brooker A, Talalay P, Fahey JW (2015) Sulforaphane improves the bronchoprotective response in asthmatics through Nrf2-mediated gene pathways. Respir Res 16:106

Calcabrini C, Maffei F, Turrini E, Fimognari C (2020) Sulforaphane potentiates anticancer effects of doxorubicin and cisplatin and mitigates their toxic effects. Front Pharmacol 11:567

Cannon LM, Butler FN, Wan W, Zhou ZS (2002) A stereospecific colorimetric assay for (S,S)-adenosylmethionine quantification based on thiopurine methyltransferase-catalyzed thiol methylation. Anal Biochem 308(2):358–363

Cascajosa-Lira A, Prieto AI, Pichardo S, Jos A, Cameán AM (2024) Protective effects of sulforaphane against toxic substances and contaminants: a systematic review. Phytomedicine 130:155731

Cazalla E, Cuadrado A, García-Yagüe ÁJ (2024) Role of the transcription factor NRF2 in maintaining the integrity of the blood-brain barrier. Fluids Barriers CNS 21:1–29

Christiansen B, Bellostas Muguerza N, Petersen AM, Kveiborg B, Madsen CR, Thomas H, Dragsted LO (2010) Ingestion of broccoli sprouts does not improve endothelial function in humans with hypertension. PLoS One 5(8):e12461

Clarke JD, Hsu A, Williams DE, Dashwood RH, Stevens JF, Yamamoto M, Ho E (2011) Metabolism and tissue distribution of sulforaphane in Nrf2 knockout and wild-type mice. Pharm Res 28(12):3171–3179

Conzatti A, da Silva Fróes FCT, Perry IDS, de Souza CG (2015) Clinical and molecular evidence of the consumption of broccoli, glucoraphanin and sulforaphane in humans. Nutr Hosp 31(2):559–569

Cook IT (2011) Analysis of the structural and kinetic properties of human SULT2A1 induced by the binding of 3′-phosphoadenosine-5′-phosphosulfate. Dissertation, University of Alabama at Birmingham

Cramer JM, Teran-Garcia M, Jeffery EH (2012) Enhancing sulforaphane absorption and excretion in healthy men through the combined consumption of fresh broccoli sprouts and a glucoraphanin-rich powder. Br J Nutr 107(9):1333–1338

Czerwiec K, Myślińska D, Wądołowska A, Ruciński J, Podlacha M, Kosiński A, Grzybiak M (2016) Charakterystyka mediatorów zapalenia—rola cytokin prozapalnych. Ann Acad Medicae Gedanensis 46(1):53–58

de Figueiredo S, Binda N, Nogueira-Machado J, Vieira-Filho S, Caligiorne R (2015) The antioxidant properties of organosulfur compounds (sulforaphane). Recent Pat Endocr Metab Immune Drug Discov 9:24–39

Desai P, Thakkar A, Ann D, Wang J, Prabhu S (2019) Loratadine self-microemulsifying drug delivery systems (SMEDDS) in combination with sulforaphane for the synergistic chemoprevention of pancreatic cancer. Drug Deliv Transl Res 9:641–651

Dinkova-Kostova A, Talalay P (1999) Relation of structure of curcumin analogs to their potencies as inducers of phase II detoxication enzymes. Carcinogenesis 20:911–914

Dinkova-Kostova AT, Holtzclaw WD, Cole RN, Itoh K, Wakabayashi N, Katoh Y (2002) Direct evidence that sulfhydryl groups of Keap1 are the sensors regulating induction of phase 2 enzymes that protect against carcinogens and oxidants. Proc Natl Acad Sci USA 99:11908–11913

Dmytriv TR, Lushchak O, Lushchak VI (2025) Glucoraphanin conversion into sulforaphane and related compounds by gut microbiota. Front Physiol 16:1497566

Dourado DF, Fernandes PA, Ramos MJ (2010) Glutathione transferase classes alpha, pi, and mu: GSH activation mechanism. J Phys Chem B 114(40):12972–12980

DrugBank (n.d.) Sulforaphane DB12422. https://go.drugbank.com/drugs/DB12422. Accessed 21 Jan 2025

Edmondson DE, Mattevi A, Binda C, Li M, Hubalek F (2004) Structure and mechanism of monoamine oxidase. Curr Med Chem 11(15):1983–1993

Egner PA, Chen JG, Wang JB, Wu Y, Sun Y, Lu JH, Kensler TW (2011) Bioavailability of sulforaphane from two broccoli sprout beverages: results of a short-term, cross-over clinical trial in Qidong, China. Cancer Prev Res 4(3):384–395

Ekins S, Wrighton SA (1999) The role of CYP2B6 in human xenobiotic metabolism. Drug Metab Rev 31(3):719–754

Elbarbry F, Elrody N (2011) Potential health benefits of sulforaphane: a review of the experimental, clinical and epidemiological evidences and underlying mechanisms. J Med Plants Res 5(4):473–484

ElKhalifa D, Al-Ziftawi N, Awaisu A, Alali F, Khalil A (2023) Efficacy and tolerability of sulforaphane in the therapeutic management of cancers: a systematic review of randomized controlled trials. Front Oncol 13:1251895

Fahey JW, Talalay P (1999) Antioxidant functions of sulforaphane: a potent inducer of phase II detoxication enzymes. Food Chem Toxicol 37:973–979. https://doi.org/10.1016/S0278-6915(99)00082-4

Fahey JW, Stephenson KK, Wade KL, Talalay P (2013) Urease from Helicobacter pylori is inactivated by sulforaphane and other isothiocyanates. Biochem Biophys Res Commun 435(1):1–7

Feng Z, Wang T, Sun Y, Chen S, Hao H, Du W, Pang Y (2023) Sulforaphane suppresses paraquat-induced oxidative damage in bovine in vitro-matured oocytes through Nrf2 transduction pathway. Ecotoxicol Environ Saf 254:114747

Fimognari C, Hrelia P (2007) Sulforaphane as a promising molecule for fighting cancer. Mutat Res 635(2–3):90–104. https://doi.org/10.1016/j.mrrev.2006.10.004

Fretland AJ, Omiecinski CJ (2000) Epoxide hydrolases: biochemistry and molecular biology. Chem Biol Interact 129(1–2):41–59
Gan L, Wang W, Jiang J, Tian K, Liu W, Cao Z (2024) Dual role of Nrf2 signaling in hepatocellular carcinoma: promoting development, immune evasion, and therapeutic challenges. Front Immunol 15:1429836
Gao L, Du F, Wang J, Zhao Y, Liu J, Cai D, Zhang S (2021) Examination of the differences between sulforaphane and sulforaphene in colon cancer: a study based on next generation sequencing. Oncol Lett 22(4):1–10
Gasper AV, Al-Janobi A, Smith JA, Bacon JR, Fortun P, Atherton C, Mithen RF (2005) Glutathione S-transferase M1 polymorphism and metabolism of sulforaphane from standard and high-glucosinolate broccoli. Am J Clin Nutr 82:1283–1291
Giacoppo S, Galuppo M, Montaut S, Iori R, Rollin P, Bramanti P, Mazzon E (2015) An overview on neuroprotective effects of isothiocyanates for the treatment of neurodegenerative diseases. Fitoterapia 106:12–21
Go RE, Hwang KA, Choi KC (2015) Cytochrome P450 1 family and cancers. J Steroid Biochem Mol Biol 147:24–30
Gong TT, Liu XD, Zhan ZP, Wu QJ (2020) Sulforaphane enhances the cisplatin sensitivity through regulating DNA repair and accumulation of intracellular cisplatin in ovarian cancer cells. Exp Cell Res 393(2):112061
Grady RS, Traustadóttir T, Lagalante AF, Eggler AL (2023) Bioavailable sulforaphane quantitation in plasma by LC–MS/MS is enhanced by blocking thiols. J Agric Food Chem 71(34):12875–12882
Guerrero-Beltrán CE, Calderón-Oliver M, Pedraza-Chaverri J, Chirino YI (2012) Protective effect of sulforaphane against oxidative stress: recent advances. Exp Toxicol Pathol 64(5):503–508
Harvey CJ, Thimmulappa RK, Sethi S, Kong X, Yarmus L, Brown RH, Biswal S (2011) Targeting Nrf2 signaling improves bacterial clearance by alveolar macrophages in patients with COPD and in a mouse model. Sci Transl Med 3(78):78ra32
Hauder J, Winkler S, Bub A, Rüfer CE, Pignitter M, Somoza V (2011) LC-MS/MS quantification of sulforaphane and indole-3-carbinol metabolites in human plasma and urine after dietary intake of selenium-fortified broccoli. J Agric Food Chem 59(15):8047–8057
Hein DW, Doll MA, Fretland AJ, Leff MA, Webb SJ, Xiao GH, Feng Y (2000) Molecular genetics and epidemiology of the NAT1 and NAT2 acetylation polymorphisms. Cancer Epidemiol Biomarkers Prev 9(1):29–42
Iyanagi T (2007) Molecular mechanism of phase I and phase II drug-metabolizing enzymes: implications for detoxification. Int Rev Cytol 260:35–112
Kamal MM, Akter S, Lin CN, Nazzal S (2020) Sulforaphane as an anticancer molecule: mechanisms of action, synergistic effects, enhancement of drug safety, and delivery systems. Arch Pharm Res 43(4):371–384
Karpel E (2001) Mediatory ogólnoustrojowej odpowiedzi zapalnej—znaczenie w praktyce klinicznej intensywnej terapii. Anestezjol Intensywna Ter 3:181–190
Kassahun K, Davis M, Hu P, Martin B, Baillie T (1997) Biotransformation of the naturally occurring isothiocyanate sulforaphane in the rat: identification of phase I metabolites and glutathione conjugates. Chem Res Toxicol 10(11):1228–1233
Kerr C, Adhikary G, Grun D, George N, Eckert RL (2017) Combination cisplatin and sulforaphane treatment reduces proliferation, invasion and tumor formation in epidermal squamous cell carcinoma. Mol Carcinog 57(1):3
Keum YS, Yu S, Chang PPJ, Yuan X, Kim JH, Xu C, Kong ANT (2006) Mechanism of action of sulforaphane: inhibition of p38 mitogen-activated protein kinase isoforms contributing to the induction of antioxidant response element–mediated heme oxygenase-1 in human hepatoma HepG2 cells. Cancer Res 66(17):8804–8813
Khachik F, Bertram JS, Huang MT, Fahey JW, Talalay P (1999) Dietary carotenoids and their metabolites as potentially useful chemoprotective agents against cancer. In: Antioxidant food supplements in human health. Academic, pp 203–229

Khatoon E, Banik K, Harsha C, Sailo BL, Thakur KK, Khwairakpam AD, Kunnumakkara AB (2022) Phytochemicals in cancer cell chemosensitization: current knowledge and future perspectives. Semin Cancer Biol 80:306–339. https://doi.org/10.1016/j.semcancer.2020.06.014

Kim JH, Cheong HS, Park BL, Kim LH, Shin HJ, Na HS, Shin HD (2015) Direct sequencing and comprehensive screening of genetic polymorphisms on CYP2 family genes (CYP2A6, CYP2B6, CYP2C8, and CYP2E1) in five ethnic populations. Arch Pharm Res 38(1):115–128

Lagunas-Rangel FA (2024) Sulforaphane target protein prediction: a bioinformatics analysis. Appl Sci 14(3):1052

Langston-Cox A, Anderson D, Creek DJ, Palmer K, Wallace EM, Marshall SA (2020) Measuring sulforaphane and its metabolites in human plasma: a high throughput method. Molecules 25(4):829

Leone A, Diorio G, Sexton W, Schell M, Alexandrow M, Fahey JW, Kumar NB (2017) Sulforaphane for the chemoprevention of bladder cancer: molecular mechanism targeted approach. Oncotarget 8(21):35412

Li AP, Kaminski DL, Rasmussen A (1995) Substrates of human hepatic cytochrome P450 3A4. Toxicology 104(1–3):1–8

Li F, Hullar MA, Beresford SA, Lampe JW (2011) Variation of glucoraphanin metabolism in vivo and ex vivo by human gut bacteria. Br J Nutr 106(3):408–416

Li Y, Meeran SM, Tollefsbol TO (2017) Combinatorial bioactive botanicals re-sensitize tamoxifen treatment in ER-negative breast cancer via epigenetic reactivation of ERα expression. Sci Rep 7(1):9345

Liang J, Mantelos A, Toh ZQ, Tortorella SM, Ververis K, Vongsvivut J, Karagiannis TC (2020) Investigation of potential anti-pneumococcal effects of l-sulforaphane and metabolites: insights from synchrotron-FTIR microspectroscopy and molecular docking studies. J Mol Graph Model 97:107568

Licznerska B, Szaefer H, Krajka-Kuźniak V (2021) R-sulforaphane modulates the expression profile of AhR, ERα, Nrf2, NQO1, and GSTP in human breast cell lines. Mol Cell Biochem 476:525–533. https://doi.org/10.1007/s11010-020-03913-5

Liska DJ (1998) The detoxification enzyme systems. Altern Med Rev 3(3):187–198

Livingstone TL (2022) Accumulation of Sulphur-containing dietary bioactives and the impact on the transcriptional signature of the prostate. Doctoral dissertation, University of East Anglia. https://ueaeprints.uea.ac.uk/id/eprint/84855/

Luang-In V, Narbad A, Nueno-Palop C, Mithen R, Bennett M, Rossiter JT (2014) The metabolism of methylsulfinylalkyl- and methylthioalkyl-glucosinolates by a selection of human gut bacteria. Mol Nutr Food Res 58(4):875–883

Lubelska K, Milczarek M, Modzelewska K, Krzysztoń-Russjan J, Fronczyk K, Wiktorska K (2012) Interactions between drugs and sulforaphane modulate the drug metabolism enzymatic system. Pharmacol Rep 64(5):1243–1252

Mahéo K, Morel F, Langouët S, Kramer H, Le Ferrec E, Ketterer B, Guillouzo A (1997) Inhibition of cytochromes P-450 and induction of glutathione S-transferases by sulforaphane in primary human and rat hepatocytes. Cancer Res 57(17):3649–3652

Mahn A, Castillo A (2021) Potential of sulforaphane as a natural immune system enhancer: a review. Molecules 26(3):752

Mangla B, Javed S, Sultan MH, Kumar P, Kohli K, Najmi A, Ahsan W (2021) Sulforaphane: a review of its therapeutic potentials, advances in its nanodelivery, recent patents, and clinical trials. Phytother Res 35(10):5440–5458

Marshall SA, Young RB, Lewis JM, Rutten EL, Gould J, Barlow CK, Gulliver EL (2023) The broccoli-derived antioxidant sulforaphane changes the growth of gastrointestinal microbiota, allowing for the production of anti-inflammatory metabolites. J Funct Foods 107:105645

Matusheski NV, Juvik JA, Jeffery EH (2004) Heating decreases epithiospecifier protein activity and increases sulforaphane formation in broccoli. Phytochemistry 65(9):1273–1281

Mazarakis N, Snibson K, Licciardi PV, Karagiannis TC (2020) The potential use of l-sulforaphane for the treatment of chronic inflammatory diseases: a review of the clinical evidence. Clin Nutr 39(3):664–675

Men X, Han X, Oh G, Im JH, Lim JS, Cho GH, Lee OH (2023) Plant sources, extraction techniques, analytical methods, bioactivity, and bioavailability of sulforaphane: a review. Food Sci Biotechnol 33(3):539

Moens U, Kostenko S, Sveinbjørnsson B (2013) The role of mitogen-activated protein kinase-activated protein kinases (MAPKAPKs) in inflammation. Genes 4(2):101–133

Mohanty S, Pal A, Konkimalla VB, Sudam Chandra SI (2018) Anti-inflammatory and anti-granuloma activity of sulforaphane, a naturally occurring isothiocyanate from broccoli (*Brassica oleracea*). Asian J Pharm Clin Res 11(6):411–416

Moi P, Chan K, Asunis I, Cao A, Kan Y (1994) Isolation of NF-E2-related factor 2 (Nrf2), a NFE2-like basic leucine zipper transcriptional activator that binds to the tandem NF-E2/AP1 repeat of the beta-globin locus control region. Proc Natl Acad Sci USA 91(21):9926–9930

Mokkawes T, de Visser SP (2023) Melatonin activation by cytochrome P450 isozymes: how does CYP1A2 compare to CYP1A1? Int J Mol Sci 24(4):3651

Montgomery M, Srinivasan A (2019) Epigenetic gene regulation by dietary compounds in cancer prevention. Adv Nutr 10(6):1012–1028

Múnera-Rodríguez AM, Leiva-Castro C, Sobrino F, López-Enríquez S, Palomares F (2024) Sulforaphane-mediated immune regulation through inhibition of NF-kB and MAPK signaling pathways in human dendritic cells. Biomed Pharmacother 177:117056

Pańczyszyn-Trzewik P, Surowiec A, Wnorowski J, Zavatskyi R, Sowa-Kućma M (2020) Rola czynnika Nrf2 i jego aktywatorów w patogenezie wybranych chorób związanych z dysfunkcją układu nerwowego. Postępy Biol Komórki 47(2):101–118

Parfenova H, Liu J, Hoover DT, Fedinec AL (2020) Vasodilator effects of sulforaphane in cerebral circulation: a critical role of endogenously produced hydrogen sulfide and arteriolar smooth muscle KATP and BK channels in the brain. J Cereb Blood Flow Metab 40(10):1987–1996

Peng X, Zhou Y, Tian H, Yang G, Li C, Geng Y, Wu W (2015) Sulforaphane inhibits invasion by phosphorylating ERK1/2 to regulate E-cadherin and CD44v6 in human prostate cancer DU145 cells. Oncol Rep 34(3):1565–1572

Petri N, Tannergren C, Holst B, Mellon FA, Bao Y, Plumb GW, Williamson G (2003) Absorption/metabolism of sulforaphane and quercetin, and regulation of phase II enzymes, in human jejunum in vivo. Drug Metab Dispos 31(6):805–813

Piska K, Pękala E (2015) Enzymy i reakcje pierwszej fazy biotransformacji leków spoza układu cytochromu P-450. Farm Pol 71(8):529–534

Platz S, Piberger AL, Budnowski J, Herz C, Schreiner M, Blaut M, Rohn S (2015) Bioavailability and biotransformation of sulforaphane and erucin metabolites in different biological matrices determined by LC–MS–MS. Anal Bioanal Chem 407(7):1819–1829. https://doi.org/10.1007/s00216-015-8482-z

Qi T, Xu F, Yan X, Li S, Li H (2016) Sulforaphane exerts anti-inflammatory effects against lipopolysaccharide-induced acute lung injury in mice through the Nrf2/ARE pathway. Int J Mol Med 37(1):182–188

Radominska-Pandya A, Czernik PJ, Little JM, Battaglia E, Mackenzie PI (1999) Structural and functional studies of UDP-glucuronosyltransferases. Drug Metab Rev 31(4):817–899

Razis AF, Iori R, Ioannides C (2011) The natural chemopreventive phytochemical R-sulforaphane is a far more potent inducer of the carcinogen-detoxifying enzyme systems in rat liver and lung than the S-isomer. Int J Cancer 128(12):2775–2782. https://doi.org/10.1002/ijc.25620

Riedl MA, Saxon A, Diaz-Sanchez D (2009) Oral sulforaphane increases phase II antioxidant enzymes in the human upper airway. Clin Immunol 130(3):244–251

Ross IA (2024) Cancer of the urinary bladder. In: Plant-based therapeutics, volume 2: the Brassicaceae family. Springer, Cham, pp 535–586

Ruhee RT, Suzuki K (2024) The immunomodulatory effects of sulforaphane in exercise-induced inflammation and oxidative stress: a prospective nutraceutical. Int J Mol Sci 25(3):1790

Sabio G, Davis RJ (2014) TNF and MAP kinase signalling pathways. Semin Immunol 26(3):237–245

Saha S, Hollands W, Teucher B, Needs PW, Narbad A, Ortori CA, Kroon PA (2012) Isothiocyanate concentrations and interconversion of sulforaphane to erucin in human subjects after consumption of commercial frozen broccoli compared to fresh broccoli. Mol Nutr Food Res 56(12):1906–1916

Sajak J, Wojnarowski K, Strzałka-Mrozik B (2023) Substancje o potencjale terapeutycznym wpływające na szlak Keap1-Nrf2-ARE w chorobach oczu związanych z wiekiem. Farm Pol 99(4):237–248

Schepici G, Bramanti P, Mazzon E (2020) Efficacy of sulforaphane in neurodegenerative diseases. Int J 21

Shapiro TA, Fahey JW, Wade KL, Stephenson KK, Talalay P (1998) Human metabolism and excretion of cancer chemoprotective glucosinolates and isothiocyanates of cruciferous vegetables. Cancer Epidemiol Biomarkers Prev 7(12):1091–1100

Sharma C, Sadrieh L, Priyani A, Ahmed M, Hassan AH, Hussain A (2011) Anti-carcinogenic effects of sulforaphane in association with its apoptosis-inducing and anti-inflammatory properties in human cervical cancer cells. Cancer Epidemiol 35(3):272–278

Shekarri Q, Dekker M (2021) A physiological-based model for simulating the bioavailability and kinetics of sulforaphane from broccoli products. Foods 10(11):2761

Siegel D, Gustafson DL, Dehn DL, Han JY, Boonchoong P, Berliner LJ, Ross D (2004) NAD(P) H: quinone oxidoreductase 1: role as a superoxide scavenger. Mol Pharmacol 65(5):1238–1247

Sikorska-Zimny K, Badełek E (2014) The influence of time and storage condition on level of indole-3-carbinol and sulphoraphane in Brussels sprout. In: 3rd international conference on effects of pre- and post-harvest factors on health promoting components and quality of horticultural commodities, Skierniewice, Poland, 23–25 March, Instytut Ogrodnictwa

Sikorska-Zimny K, Beneduce L (2021) The glucosinolates and their bioactive derivatives in Brassica: a review on classification, biosynthesis and content in plant tissues, fate during and after processing, effect on the human organism and interaction with the gut microbiota. Crit Rev Food Sci Nutr 61(15):2544–2571

Son ES, Fei X, Yoon JH, Seo SY, Maeng HJ, Jeong SH, Kim YC (2021) Comparison of pharmacokinetics and anti-pulmonary fibrosis-related effects of sulforaphane and sulforaphane N-acetylcysteine. Pharmaceutics 13(7):958

Srovnalova A, Vanduchova A, Svecarova M, Anzenbacherova E, Tomankova V, Anzenbacher P, Dvorak Z (2015) Effects of sulforaphane and its S- and R-enantiomers on the expression and activities of human drug-metabolizing cytochromes P450. J Funct Foods 14:487–501

Stępkowski TM, Kruszewski MK (2011) Molecular cross-talk between the NRF2/KEAP1 signaling pathway, autophagy, and apoptosis. Free Radic Biol Med 50(9):1186–1195

Strolin Benedetti M, Whomsley R, Baltes E (2006) Involvement of enzymes other than CYPs in the oxidative metabolism of xenobiotics. Expert Opin Drug Metab Toxicol 2(6):895–921

Su X, Jiang X, Meng L, Dong X, Shen Y, Xin Y (2018) Anticancer activity of sulforaphane: the epigenetic mechanisms and the Nrf2 signaling pathway. Oxidative Med Cell Longev 2018:5438179

Subedi L, Lee JH, Yumnam S, Ji E, Kim SY (2019) Anti-inflammatory effect of sulforaphane on LPS-activated microglia potentially through JNK/AP-1/NF-κB inhibition and Nrf2/HO-1 activation. Cells 8(2):194

Sun Y, Yang T, Mao L, Zhang F (2017) Sulforaphane protects against brain diseases: roles of cytoprotective enzymes. Austin J Cerebrovasc Dis Stroke 4(1)

Sutaria D, Grandhi BK, Thakkar A, Wang J, Prabhu S (2012) Chemoprevention of pancreatic cancer using solid-lipid nanoparticulate delivery of a novel aspirin, curcumin and sulforaphane drug combination regimen. Int J Oncol 41(6):2260–2268

Talalay P, Fahey JW, Holtzclaw WD, Prestera T, Zhang Y (1995) Chemoprotection against cancer by phase 2 enzyme induction. Toxicol Lett 82:173–179. https://doi.org/10.1016/0378-4274(95)03553-2

The Human Metabolome Database (n.d.) Metabocard for sulforaphane (HMDB0005792) and sulforaphane-cysteine-glycine (HMDB0240559). https://hmdb.ca/metabolites/. Accessed 28 Jan 2025

Thiruvengadam M, Venkidasamy B, Subramanian U, Samynathan R, Ali Shariati M, Rebezov M, Chung IM (2021) Bioactive compounds in oxidative stress-mediated diseases: targeting the NRF2/ARE signaling pathway and epigenetic regulation. Antioxidants 10(12):1859

Tortorella SM, Royce SG, Licciardi PV, Karagiannis TC (2015) Dietary sulforaphane in cancer chemoprevention: the role of epigenetic regulation and HDAC inhibition. Antioxid Redox Signal 22(16):1382–1424

Townsend BE, Johnson RW (2016) Sulforaphane induces Nrf2 target genes and attenuates inflammatory gene expression in microglia from brain of young adult and aged mice. Exp Gerontol 73:42–48

Van Steenwijk HP, Winter E, Knaven E, Brouwers JF, Van Baardwijk M, Van Dalum JB, De Boer A (2023) The beneficial effect of sulforaphane on platelet responsiveness during caloric load: a single-intake, double-blind, placebo-controlled, crossover trial in healthy participants. Front Nutr 10:1204561

Vanduchova A, Tomankova V, Anzenbacher P, Anzenbacherova E (2016) Influence of sulforaphane metabolites on activities of human drug-metabolizing cytochrome P450 and determination of sulforaphane in human liver cells. J Med Food 19(12):1141–1146

Venkatakrishnan K, Greenblatt DJ, von Moltke LL, Schmider J, Harmatz JS, Shader RI (1998) Five distinct human cytochromes mediate amitriptyline N-demethylation in vitro: dominance of CYP 2C19 and 3A4. J Clin Pharmacol 38(2):112–121

Vergara F, Wenzler M, Hansen BG, Kliebenstein DJ, Halkier BA, Gershenzon J, Schneider B (2008) Determination of the absolute configuration of the glucosinolate methyl sulfoxide group reveals a stereospecific biosynthesis of the side chain. Phytochemistry 69(15):2737–2742

Vu GH, Nguyen HD (2024) Molecular mechanisms of sulforaphane in Alzheimer's disease: insights from an in-silico study. In Silico Pharmacol 12(2):96

Wang M, Zhu JY, Chen S, Qing Y, Wu D, Lin YM, Li YQ (2014) Effects of co-treatment with sulforaphane and autophagy modulators on UGT1A isoforms and cytochrome P450 3A4 expression in Caco-2 human colon cancer cells. Oncol Lett 8(6):2407–2416

Wang F, Wang W, Li J, Zhang J, Wang X, Wang M (2018) Sulforaphane reverses gefitinib tolerance in human lung cancer cells via modulation of sonic hedgehog signaling. Oncol Lett 15(1):109–114

Wang F, Sun Y, Huang X, Qiao C, Zhang W, Liu P, Wang M (2021) Sulforaphane inhibits self-renewal of lung cancer stem cells through the modulation of sonic hedgehog signaling pathway and polyhomeotic homolog 3. AMB Express 11:1–10

Wogan GN, Hecht SS, Felton JS, Conney AH, Loeb LA (2004) Environmental and chemical carcinogenesis. Semin Cancer Biol 14(6):473–486

Xie H, Chun FKH, Rutz J, Blaheta RA (2021) Sulforaphane impact on reactive oxygen species (ROS) in bladder carcinoma. Int J Mol Sci 22(11):5938

Yagishita Y, Fahey JW, Dinkova-Kostova AT, Kensler TW (2019) Broccoli or sulforaphane: is it the source or dose that matters? Molecules 24(19):3593

Yan L, Yan Y (2023) Therapeutic potential of sulforaphane in liver diseases: a review. Front Pharmacol 14:1256029

Yanaka A, Fahey JW, Fukumoto A, Nakayama M, Inoue S, Zhang S, Yamamoto M (2009) Dietary sulforaphane-rich broccoli sprouts reduce colonization and attenuate gastritis in Helicobacter pylori-infected mice and humans. Cancer Prev Res 2(4):353–360

Yang L, Palliyaguru DL, Kensler TW (2016) Frugal chemoprevention: targeting Nrf2 with foods rich in sulforaphane. Semin Oncol 43(1):146–153

Yoxall V, Kentish P, Coldham N, Kuhnert N, Sauer MJ, Ioannides C (2005) Modulation of hepatic cytochromes P450 and phase II enzymes by dietary doses of sulforaphane in rats: implications for its chemopreventive activity. Int J Cancer 117(3):356–362

Yu A, Dong H, Lang D, Haining RL (2001) Characterization of dextromethorphan O- and N-demethylation catalyzed by highly purified recombinant human CYP2D6. Drug Metab Dispos 29(11):1362–1365

Yuan B, Liu Y, Yu X, Yin L, Peng Y, Gao Y, Li X (2018) FOXM1 contributes to taxane resistance by regulating UHRF1-controlled cancer cell stemness. Cell Death Dis 9(5):562

Zamora-Díaz IY, González-Trujano ME, Martínez-Vargas D, Moreno-Pérez GF, Hernandez-Leon A, Narváez-González HF, López-Muñoz FJ (2025) Pharmacological interactions of sulforaphane and gabapentin in a murine fibromyalgia-like pain model. Biomed Pharmacother 184:117929

Zhang Y, Callaway EC (2002) High cellular accumulation of sulphoraphane, a dietary anticarcinogen, is followed by rapid transporter-mediated export as a glutathione conjugate. Biochem J 364(1):301–307

Zhang R, Zhang J, Fang L, Li X, Zhao Y, Shi W, An L (2014) Neuroprotective effects of sulforaphane on cholinergic neurons in mice with Alzheimer's disease-like lesions. Int J Mol Sci 15(8):14396–14410

Zhang Z, Atwell LL, Farris PE, Ho E, Shannon J (2016) Associations between cruciferous vegetable intake and selected biomarkers among women scheduled for breast biopsies. Public Health Nutr 19(7):1288–1295

Zhang Y, Wang Z, Wang Y, Jin W, Zhang Z, Jin L, Zheng L (2024) CYP3A4 and CYP3A5: the crucial roles in clinical drug metabolism and the significant implications of genetic polymorphisms. PeerJ 12:e18636

Zhang W, Ren H, Chen W, Hu B, Feng C, Li P, Fang J (2025) Nicotinamide phosphoribosyltransferase in NAD+ metabolism: physiological and pathophysiological implications. Cell Death Discov 11(1):371

Zhao Z, Chen Q, Qiao X, Wang J, Ali ZTA, Li J, Yin L (2025) Sulforaphane in cancer precision medicine: from biosynthetic origins to multiscale mechanisms and clinical translation. Front Immunol 16:1702860

Zhou SF, Zhou ZW, Yang LP, Cai JP (2009) Substrates, inducers, inhibitors and structure-activity relationships of human cytochrome P450 2C9 and implications in drug development. Curr Med Chem 16(27):3480–3675

Zuo M, Chen H, Liao Y, He P, Xu T, Tang J, Zhang N (2023) Sulforaphane and bladder cancer: a potential novel antitumor compound. Front Pharmacol 14:1254236

Chapter 3
Therapeutic Applications of Sulforaphane in Human Disease

Abstract Sulforaphane (SFN) demonstrates broad and multifaceted therapeutic potential across oncological, neurological, cardiovascular, metabolic, inflammatory, and infectious disease contexts. In cancer models, SFN targets hallmarks of malignancy through epigenetic reprogramming via HDAC and DNMT inhibition, modulation of the hypoxic tumour microenvironment, induction of apoptosis through Bcl-2 family rebalancing, and suppression of invasion via MMP-2, CA IX, and CD44v6 downregulation. Neuroprotective effects—mediated through Nrf2/ARE activation, microglial M1-to-M2 polarisation, and blood–brain barrier preservation—have been demonstrated in models of Alzheimer's and Parkinson's disease, peripheral neuropathy, schizophrenia, and autism spectrum disorder. Cardioprotective activity encompasses reduction of ischaemia-reperfusion injury, attenuation of mitochondrial and diastolic dysfunction in the diabetic heart, anti-thrombotic effects, and reversal of leptin resistance. Anti-inflammatory and immunomodulatory properties are primarily mediated through NF-κB and NLRP3 inflammasome inhibition, Treg proliferation, and epigenetic suppression of inflammatory gene expression, with proposed utility as a steroid-sparing agent. Antimicrobial activity spans Gram-positive and Gram-negative bacteria—including multidrug-resistant strains—as well as *Candida* species, operating through membrane disruption, biofilm inhibition, and quorum sensing interference. A critical translational consideration throughout is the concentration-dependence of observed effects: findings are evaluated across physiological, intermediate, and supraphysiological tiers to assess their clinical relevance and dietary attainability.

Keywords Therapeutic potential · Cancer · Neuroprotection · Anti-inflammatory effects · Antimicrobial activity

K. Sikorska-Zimny, *Sulforaphane in Human Health*, SpringerBriefs in Food, Health, and Nutrition, https://doi.org/10.1007/978-3-032-34331-4_3

3.1 Therapeutic Potential of Sulforaphane

The biological activity of sulforaphane in the human body is primarily associated with its potent anti-inflammatory properties, which serve as the foundation for its therapeutic potential. The following is a review of the most recent and promising findings regarding biomedical applications (Table 3.1). The biological effects of sulforaphane in humans include, but are not limited to, the following:

- Anti-apoptotic and anti-cancer effects (Russo et al. 2018; Calcabrini et al. 2020).
- Influencing: neurological disorders/nervous system (Huang et al. 2019; Uddin et al. 2020), diabetes-related diseases (Li et al. 2019; Sun et al. 2020), and cardio muscular disease (Xin et al. 2018; Wang et al. 2022).
- Preventing inflammatory degenerative diseases (Schepici et al. 2020; Calabrese and Kozumbo 2021) therapeutically effect in an atopic dermatitis, psoriasis (Wu et al. 2019; Ma et al. 2024).
- Reduce/fight obese (Martins et al. 2018; Çakır et al. 2022).
- Anti-arthritis effects (Moon et al. 2021; Ma et al. 2024).
- Potentially attenuates arsenic-, chromium-, aluminium-, cadmium-induced toxicity (Ogunlade et al. 2020; Yang et al. 2020; Baralić et al. 2023; Cascajosa-Lira et al. 2024).

The available experimental studies of SFN in vitro show differences in the concentrations of SFN used. Reported values range from 0.625 μM (Gao et al. 2018) to 200 μM (Clulow et al. 2017). Human bioavailability studies have indicated that the plasma concentrations of SFN after dietary intake are relatively low. Following the consumption of broccoli sprouts providing approximately 100–200 μmol SFN equivalent, peak plasma levels typically reach 1–2 μM to a maximum of 2.27 μM/L of ITC plasma concentration with a dose of 200 mmol of broccoli sprouts (Clarke et al. 2008; Atwell et al. 2015). The consumption of super broccoli soup showed C_{max} 7.3 μM compared to 2.2 μM for regular ("*The super broccoli soups contained 3.4-fold greater amounts of sulforaphane and sulforaphane-derived thiol conjugates than did the standard broccoli soups*") (Gasper et al. 2005).

Sulforaphane shows clinical efficacy in humans at doses as low as 0.5 μmol/kg, with typical trial doses averaging 100 μmol for sulforaphane and 190 μmol for its precursor glucoraphanin (Yagishita et al. 2019). As many preclinical studies exceed tolerable human levels, there is a critical need to align experimental dosages with achievable human concentrations. The integration of pharmacological limits and standardised formulations is essential for advancing broccoli-based strategies into effective public health paradigms.

Most in vitro studies use concentrations within the 5–40 μM range. The lower end of this interval (5–15 μM) partially overlaps with concentrations that may be transiently achievable following the consumption of broccoli-rich foods or supplements. However, concentrations above 40 μM clearly exceed physiologically plausible exposure levels and should, therefore, be interpreted with caution.

Table 3.1 Scientific trials on the influence of SFN on health

Disease	Target	Used form	Dose	Effect (on the examined target, cells)		Note	Author
				Decrease, down regulates (substance)/ knockdown (genes)	*Increase (substance)/ induce (process)/ up-regulate (genes)*		
Prostate cancer	Human cell lines (LNCaP and PC-3) hormone-dependent and independent lines	SNF dissolved in DMSO	5 and 15 μM conc.	Cell proliferation, HDAC,	GAPDH, MAP1LC3B2, H2AFY		Beaver et al. (2017)
Prostate cancer	DU145 and PC3 prostate cancer cells	Incubation with SFN	1–20 μM	G2/M-cells	H3, H4 histones acetylation	Altering the intracellular signalling cascade	Rutz et al. (2020)
Nasopharyngeal cancer	Human cell lines (Hone1, CNE1, CNE2, Sune1)	SNF dissolved in DMSO	5, 10, 20 μmol/L conc.	Growth and proliferation,	SFN was added at different concentrations (0, 5, 10, and 20 μmol/L) every day for 10 days	Apoptosis	Li et al. (2018)
Breast cancer (triple negative)	Human cell lines (SUM149, SUM159)	Cells treated with various concentrations of SFN	2.5 μM and 5 μM	NF-κB, protein target cyclin D1, IL-6, IL-8			Burnett et al. (2017)
Lung adenocarcinoma, embryonic kidney, colorectal adenocarcinoma	Human cell lines (A549, HEK293, HT29)	Cells treated with various concentrations of SFN	0.625 μM to 40 μM	Protein levels of DNMTs, HDACs, and its target gene CDH1			Gao et al. (2018)
Human breast cancer	Human cell lines, (MDA-MB-231, MCF-7, T47D)	SNF dissolved in DMSO	0, 5, 15, 25 μmol/L	Proteins: ER-α, EGFR and HER-2	Cells were treated with SNF for 72 h	Apoptosis	Pledgie-Tracy et al. (2007)

(continued)

Table 3.1 (continued)

Disease	Target	Used form	Dose	Effect (on the examined target, cells)		Note	Author
				Decrease, down regulates (substance)/ knockdown (genes)	*Increase (substance)/ induce (process)/ up-regulate (genes)*		
Prostate cancer	Human cell lines (DU-145)	Broccoli extracts (juice) added into cultures	0.19–100% were added to the cultures and incubated at 37 °C for 24 and 48 h in 5% CO_2	50% death of the cancer cells	Broccoli extracts at ten different doses were added to the cultures and incubated at 37 °C for 24 and 48 h in 5% CO_2		Türker et al. (2019)
Bladder cancer	Human cell lines (T24)	Double wash in SFN	Medium containing 2.5–160 μM SFN	Cell proliferation and growth (conc. 10–160 μM), GSH, UGT, COX-2	Cell proliferation and growth (conc. 2.5 μM) Nrf2–6-48 h	Treated and incubated with SFN for 6, 24, and 48 h,	He et al. (2021)
Colorectal cancer	Human cell lines (SW620)	Cells treated with SFN	10, 30, and 60 mg/L	Bcl-2, PI3K, p-Akt, p-mTOR	Inhibition rate of cell proliferation in each group increased with the extension of time	Apoptosis, Bax	Jiang et al. (2022)
Colorectal cancer	Human cell lines (HCT116)	Cells treated with SFN	30, 60, and 120 μM	Cyclooxygenase-2, protein kinase B, glycogen synthase kinase-3 beta		Apoptosis	Zhu et al. (2022)
Small-cell lung cancer	Human cell lines (NCI-H69, NCI-H69AR, NCI-H82)	SFN in DMSO	1, 5, 10, 20 μM	Cell growth, mRNA, and protein expression levels of SLC7A11 (cystine/ glutamate antiporter xCT)		Induced cell death was mediated via ferroptosis	Iida et al. (2021)

Breast cancer	Human cell lines (MCF7, MDA-MB-231, MCF10A)	R-SFN in DMSO	1, 10, 20 μM		Nrf2, *NQO1*, GSTP	R-SFN was used after a 24 h pre-incubation period	Licznerska et al. (2021)
Colorectal cancer	Human cell lines (HTC-116)	SFN in DMSO	0, 1, 2, 4, 8, 16 μM	Cell viability	Cells treated with appropriate concentration of SFN for 12, 24, 48 h	Ferroptosis of HCT-116 cells by activating the SIRT3/AMPK/ mTOR signalling pathway.	Hu et al. (2024)
Lung cancer	Human cell lines (A549, H460)	SFN added to the wells with cells	0, 2, 4, 6, 8, 10, 12 μmol/L	Proliferation and self-renewal of stem cells, protein expression of PHC3 in CD133+, activities of SHh,			Wang et al. (2021)
Lung cancer	Human cell lines (H1299, 95C, and 95D)	Cells were treated with sulforaphane	0.5, 1, 2.5, 5, 10, 25, 50, and 100 μmol/L for 48 h	Cell proliferation miR-616-5p levels The GSK3β/β--catenin signalling pathway		SFN—Potential adjuvant chemotherapeutic agent	Wang et al. (2017)
Glioblastoma	Mice	SFN dissolved in corn oil, administered intraperitoneally	20 mg/kg/d	TrxR1	Sensitise glioblastoma tumours to chemoradiotherapy	Enhancing chemo-radiotherapy, induced apoptosis	Ge et al. (2025)
Colorectal carcinoma	Human cell lines (HT-29, HT-116)	SFN in DMSO	0, 4, 8, 16 and 32 μmol/L	Total number of viable cells		Anti-cancer effect in p53 mutation-dependent manner	Đurić et al. (2023)

(continued)

Table 3.1 (continued)

Disease	Target	Used form	Dose	Effect (on the examined target, cells)		Note	Author
				Decrease, down regulates (substance)/ knockdown (genes)	*Increase (substance)/ induce (process)/ up-regulate (genes)*		
Lung adeno/ squamous carcinoma	Human cell lines (A549, SK-1)	D, L-Sulforaphane	20 μM	Polymerisation of microtubules, microtubule-mediated mitophagy, the fusion of mitophagosomes to lysosomes	Apoptosis possible from the downregulation of FA synthesis-associated proteins, the dysfunction of microtubule- and autophagy-associated membrane proteins	Apoptosis, cleaved caspase-3 level	Yan et al. (2021)
Dermal fibroblasts, prostate cancer	Human cell lines (HDFa, PC3)	D, L-sulforaphane in DMSO (a stock concentration of 2.5–40 mM)	10, 20, 40 μM		Possible protect cells against an accumulation of mHtt aggregates—a main cause of Huntington's disease	Autophagy in cancer cells	Borkowska et al. (2024)
Bronchial carcinoid cancer	Human cell lines (NCI-H727—Typical carcinoid; NCI-H720—Atypical carcinoid)	SFN in combination with acetazolamide	10, 20, 40 μM	PI3K/Akt/mTOR survival pathway, 5-HT secretion		Apoptosis	Mokhtari et al. (2022)
Monoblastic leukaemia	Human monoblastic leukaemia U937 cells (RCB0435)	Cells incubated with SFN	1,2,5 μM	Proliferation (dose manner)		Cytotoxicity	Akiyoshi et al. (2019)

Non-small cell lung cancer	Human cell lines (H1299)	Cells treated with SFN	5, 10, 15 μM	Cyclin B1 expression	Cell cycle arrest, cyclin D1, K expression		Żuryń et al. (2019)
Breast cancer	Human cell line (MDA-MB-231)	Cells treated with SFN	5, 10, 20, 30, 40 μM	Migration of cells, gene expression of ZEB1, fibronectin, and claudin-1		Apoptosis	Bagheri et al. (2020)
Gastric cancer cells	Human cell line (HGC27)	Cells treated with SFN	20, 40, 80 μg/mL	ROS, ATP, expression of Bcl-2	Expression levels of Bax and p53		LiFang et al. (2019)
Hepatocellular carcinoma, human hepatocyte line	Human cell lines (HepG2, HHL5)	Cells treated with SFN	1.25–160 μM	Cell viability	DNA strand breaks, Nrf2, GSH		Liu et al. (2019)
Breast cancer	Human cell lines (MCF-7, MDA-MB-231)	Sulforaphane in DMSO	5, 40 μM	Proliferation, the size, and number of PML NBs	Cell toxicity, accumulation of cytoplasmic PML		Alhazmi et al. (2020)
Lung, ovarian, cervical, lung, colon, prostate cancer	Human cell line (A549, CHO, HeLa, Hepa1c1c7, HT-29)	Cells cultured with SFN (0.5–16 μg/mL)	2, 4, 6, 8, 10, 12, 14, 16 μg/mL	Cell proliferation	Cell cycle arrest		Lim et al. (2020)
Acute myeloid leukaemia	Human cell lines (937, KG-1, HL-60, and NB-4	Cells incubation with SFN	15, 30, 45 and 60 μM)	Number of live cells		Mortality rate of acute myeloid leukaemia cells, apoptosis	Koolivand et al. (2018)
Lung cancer	Human cell line (H460)	Cells incubation with SFN	1, 2, 5 μmol/L	Proliferation, expression of β-catenin, Oct4, Sox2, c-Myc, Nanog		Cells apoptosis	Xie et al. (2016)

(continued)

Table 3.1 (continued)

Disease	Target	Used form	Dose	Effect (on the examined target, cells)		Note	Author
				Decrease, down regulates (substance)/ knockdown (genes)	*Increase (substance)/ induce (process)/ up-regulate (genes)*		
Breast cancer	Human cell line (MCF7 and MDA-MB-231)	Cells incubated with SFN	5, 25, 100, 200 μM	NF-κB		Cell apoptosis, IκBα	Clulow et al. (2017)
Oral squamous cell carcinoma (OSCC)	OSCC cell lines SAS and GNM	Cells incubated with SFN	10, 20, 40, 80 μM	Migration, ALDH1 activity, CD44 expression	Levels of miR-200		Liu et al. (2017)
Gastric carcinoma	Human cell lines (MGC803, BGC823)	Cells treated with SFN	2.5, 5, 10, 20, 40 μM	mRNA level of IL-6R	miR-124		Wang et al. (2016)
Hepatocellular carcinoma	Human cell line (HepG2)	Sulforaphane in DMSO	20, 40, 60 μM	Formation of the colonies	Cell migration, Nrf2		Liu et al. (2018)
Gastric cancer	Human cell line (GHC27)	Cells treated with SFN	10, 20, 40 μg/ml	Proliferation, Bcl-2, PI3K, p-Akt/Akt		Apoptotic, Bax	Zhang and Tu (2018)
Gastric carcinoma	Human cell lines (MGC803, AGS)	Cells treated with SFN	2, 8, 32 μM	Proliferation, migration, SMYD3, myosin regulatory light chain 9, and cysteine-rich angiogenic inducer 61	SMYD3 might be an important mediator of SFN (overexpression of SMYD3 could abolish the effects of SFN)	Cell cycle arrest, apoptosis	Dong et al. (2018)
Antidiabetic activity	In vitro, method with amylase enzyme usage	Sulforaphane from broccoli extract	100, 250, 500 μg	Amylase		Hypoglycaemic activity	Basha et al. (2017)

Insulin resistance	Human volunteers	Sulforaphane in broccoli sprouts powder	5, 10 g/d	Serum insulin concentration			Bahadoran et al. (2012)
Acute inflammatory (paw endema)	Rats, mice	Oral	2–5 mg/kg	IL-6, TNF-α, GSH, catalase, SOD	SGPT, SGOT, ALP		Mohanty et al. (2018)
Anti-inflammatory activity	Murine macrophage cells (RAW264.7)	Sulforaphane treatment before the induction of inflammation via LPS	2.5 μM	NF-κB activity, inflammasome activation, IL-1β		SFN may be a suitable treatment for disruption of inflammatory pathways (mediated through direct interactions with macrophage migration inhibitory factor)	Treasure et al. (2024)
Maternal endothelial dysfunction	Human umbilical vein endothelial cells (HUVECs)	Cells treated with SFN	5, 10, 20 μM	TNF-α, secretion of endothelin-1, VCAM1, ICAM1, E-selectin,		SFN prevented increased endothelial permeability, confers placental protection through mechanisms beyond the NRF2 pathway	Cox et al. (2019)
Acute lung injury/acute respiratory distress syndrome (ARDS)	Mice	Injection	50 mg/kg	LDH, IL-6, TNF-α, PGE2		Mice were treated with sulforaphane after 3 days ALI was induced by the administration of LPS	Qi et al. (2016)

(continued)

Table 3.1 (continued)

Disease	Target	Used form	Dose	Effect (on the examined target, cells)		Note	Author
				Decrease, down regulates (substance)/ knockdown (genes)	*Increase (substance)/ induce (process)/ up-regulate (genes)*		
Tuberculosis	Strains (*M. smegmatis* mc2155, *M. tuberculosis* H37Ra, and *Escherichia coli* MG1655)	Cells cultured with SFN (100 μg/ml)	50, 100, 200 μg/ml (for *M. smegmatis*) 25, 50, 100, 200 μg/ml for *M. tuberculosis* H37Ra)			Antimicrobial, bactericidal effect of macrophages on intracellular bacteria	Zhao et al. (2022)

5-HT serotonin-producing tumour, *ALDH1* aldehyde dehydrogenase 1 family, *ALP* alkaline phosphatase, *ATP* adenosine triphosphate, *CD44* an acidic cell surface adhesion protein, *CDH1* cadherin-1, *c-Myc/ MYCC/bHLHe39*, proto-oncogene, bHLH transcription factor, *COX-2* cyclooxygenase-2, *DMSO* dimethyl sulfoxide, *DNMT* DNA methyltransferase, *EGFR* epidermal growth factor receptor, *ERK* activate extracellular signal-regulated kinase, *ERα* oestrogen receptor alpha, *GAPDH* glyceraldehyde 3-phosphate dehydrogenase, *GNM* gingival carcinoma neck metastasis proliferation, *GSTP* glutathione S-transferase P, *H2AFY* a protein, in humans is encoded by the H2AFY gene, *HDAC* inhibitors and histone deacetylase, *HER2* receptor tyrosine-protein kinase erbB-2, *IL-6* interleukin-6, *LDH* lactate dehydrogenase, *MAP1LC3B2* microtubule associated protein 1 light chain 3 beta 2, *miR-124* microRNA-124, directly targets the 3′-untranslated regions (UTR) of the IL-6R and STAT3, *miR-200c* an RNA Gene and is affiliated with the miRNA class, diseases associated with MIR200C include cancer, *Nanog* homeobox protein, *NBs* nuclear bodies, *NF-Kb* nuclear factor kappa-light-chain-enhancer of activated B cells, *NQO1* NAD(P)H Quinone Dehydrogenase 1, *Nrf2* nuclear factor erythroid 2-related factor 2, *Oct4/POU5F1* homeodomain transcription factor of the POU family, *p-Akt* phosphorylated form of Akt (protein kinase B), *PGE2* prostaglandin-E2, *PI3K* phosphatidylinositol 3-kinases, *PML* promyelocytic leukaemia protein, *p-mTOR* phosphorylated mTOR, *ROS* reactive oxygen species, *SAS* tongue squamous cell carcinoma, *SGOT* serum glutamic oxaloacetic transaminase, *SGPT* glutamic pyruvic transaminase, *SHh* Sonic Hedgehog signalling pathways, *SMYD3* SET and MYND domain containing 3, histone methyltransferase, *SOD* oxidative markers superoxide dismutase, *Sox2* SRY-box transcription factor 2, *SPTLC3* palmitoyltransferase 3, *TNF-α* tumour necrosis factor-α, *TrxR1* thioredoxin reductase 1, *UDP*-glucuronosyltransferase (UGT), *ZEB1* zinc-finger E-box-binding homeobox 1

Studies employing concentrations between 0.625 and 10 μM represent the most physiologically credible experimental conditions. For example, Gao et al. (2018) investigated DNA methylation and histone deacetylase activity in lung adenocarcinoma (A549), colorectal (HT-29), and kidney (HEK293) cell lines using concentrations starting at 0.625 μM (up to 40 μM). Similarly, Treasure et al. (2024) examined the anti-inflammatory effects in RAW264.7 macrophages using 2.5 μM SFN, which approximates post-prandial plasma levels in humans. Comparable concentrations were used by Beaver et al. (2017) in prostate cancer cell lines (5 and 15 μM) and by Burnett et al. (2017) in triple-negative breast cancer models (2.5 and 5 μM), although the upper limit of 15 μM approaches the boundary of dietary achievability.

However, important methodological limitations must be considered. Cell culture media lack plasma protein binding, which reduces the free fraction of SFN in vivo. In addition, static cell culture systems do not reproduce metabolic pathways. Consequently, in vitro exposure to 2.5 μM SFN is not biochemically equivalent to exposure to the same nominal concentration in human tissues.

A large proportion of studies employ intermediate concentrations, where the biological effects are robust and the concentration–response relationships are easily detectable. These include studies on bladder cancer (He et al. 2021; 2.5–160 μM SFN), small cell lung cancer and ferroptosis (Iida et al. 2021; 1–20 μM), colorectal cancer via PI3K/Akt/mTOR inhibition (Jiang et al. 2022; 10–60 mg/L), and lung cancer stem cell biology (Wang et al. 2021; 0–12 μmol/L). The popularity of the 10–40 μM range likely reflects the clear biological responses observed at these concentrations, although achieving such levels in vivo would require extremely high dietary intake or pharmacological formulation. However, a dual dose-dependent response was observed. He et al. (2021) demonstrated in bladder cancer T24 cells that 2.5 μM SFN stimulated cell growth and activated Nrf2 signalling, but concentrations between 10 and 160 μM suppressed proliferation and depleted intracellular glutathione. Such a nonlinear response illustrates the risks associated with interpreting biological mechanisms based on a single experimental concentration.

Some studies have used supraphysiological concentrations. Clulow et al. (2017) examined NF-κB suppression in MCF-7 and MDA-MB-231 breast cancer cells using concentrations of up to 200 μM. Koolivand et al. (2018) applied 15–60 μM in acute myeloid leukaemia models, while Zhang and Tu (2018) used 10–40 μg/mL in gastric cancer cells, corresponding to approximately 56–226 μM. Liu et al. (2018) studied hepatocellular carcinoma migration and Nrf2 activation at 20–60 μM, and Alhazmi et al. (2020) investigated PML NB (promyelocytic leukaemia nuclear bodies) dynamics in breast cancer cells at 5 and 40 μM. These concentrations are more consistent with pharmacological exposure than with nutritional intake. Therefore, it may be useful for pharmaceutical purposes.

It is important to note that very high concentrations of SFN can cause cytotoxic effects, which may arise from non-specific mechanisms such as oxidative stress, membrane disruption, or metabolic overload, rather than from specific pathways often associated with SFN activity, including Nrf2 activation or histone deacetylase inhibition. The apoptosis (xenograft anti-tumorigenesis experiments) was noted at >60 μmol/kg, with LD_{50} 200–300 mg/kg (mice) (Socała et al. 2017; Yagishita et al. 2019).

Methodological factors further complicate the comparisons between studies. SFN is typically dissolved in dimethyl sulfoxide (DMSO) before being added to the culture medium (due to SFN's limited aqueous solubility). However, DMSO can affect cellular processes, including gene expression and cell membrane permeability. The comparability between studies is also limited by differences in reporting units. Some authors express SFN concentrations in mass-per-volume units (μg/mL or mg/L) rather than molar concentrations, which complicates direct comparisons.

In addition, the exposure duration varies considerably between experiments. Some studies have applied a single treatment for 24–48 h, whereas others have used repeated dosing for extended periods. Given the rapid clearance of SFN in vivo, the biological relevance of prolonged static exposure in cell cultures remains uncertain.

Among the listed examination concentrations of approximately 5–15 μM, the most pragmatic threshold range was observed. Above this level, biological effects may still occur but are unlikely to be achieved through diet alone, without pharmacological enhancement of bioavailability. Below this range, the findings have greater translational significance but are often more difficult to demonstrate experimentally.

3.2 Anti-cancer Activity

Most scientists have highlighted the ability of sulforaphane to modulate signalling pathways and regulate genes involved in the tumour cell cycle, ultimately leading to apoptosis (de Figueiredo et al. 2015). The proposed anti-cancer mechanism of SFN is based on its capacity to reverse epigenetic alterations in cancer cells through its interactions with histone deacetylases (HDACs), DNA methyltransferases (DNMTs), and non-coding RNAs (Su et al. 2018; Kaboli et al. 2020; Mangla et al. 2021). SFN inhibits tumour proliferation by inducing G2/M phase arrest, downregulating cyclin genes (B1 and D1), and increasing the expression of p21WAF1/CIP1, an inhibitor of cyclin-dependent kinases (Cheng et al. 2016; Żuryń et al. 2016; Kim et al. 2017; Su et al. 2018) (Fig. 3.1).

Under hypoxic conditions, SFN activates anti-cancer responses (p53, ARE, IRF-1, Pax-6, and XRE) while simultaneously inhibiting tumour progression pathways (AP-1 and HIF-1). Additionally, SFN reduces HIF-1α protein levels without affecting its stability or transcription (Kim et al. 2015; Pastorek et al. 2015). SFN also induces tumour cell apoptosis by modulating the expression of key regulatory proteins, downregulating anti-apoptotic Bcl-2 and Bcl-xL, and upregulating the pro-apoptotic protein Bax (Kim et al. 2017).

Furthermore, SFN impairs tumour cell migration and invasion by suppressing the transcription and protein expression of carbonic anhydrase IX (CA IX), an HIF-1 target involved in protecting tumour cells from hypoxia-induced pH imbalance (Pastorek et al. 2015). Peng et al. (2015) further reported that SFN inhibits tumour progression by activating ERK1/2, which upregulates E-cadherin (a suppressor of invasion) and downregulates CD44v6, ultimately reducing the expression

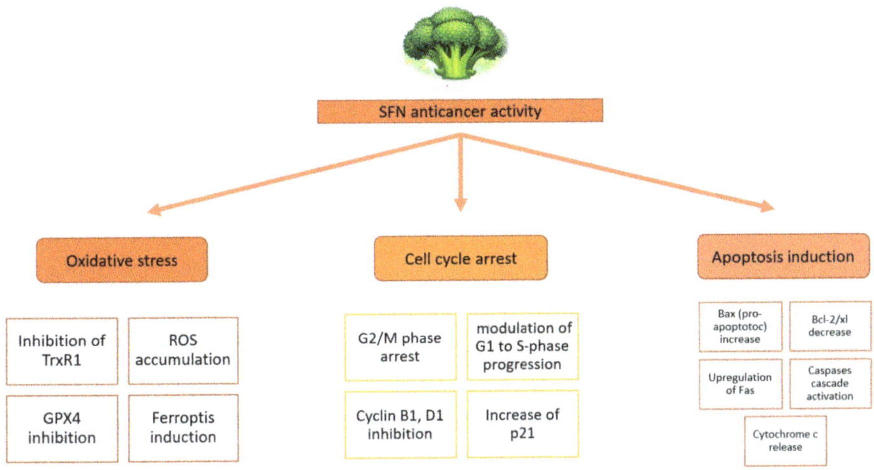

Fig. 3.1 Anti-cancer mechanisms of sulforaphane

and activity of matrix metalloproteinase-2 (MMP-2), a key invasion-promoting factor. Detailed investigations have demonstrated that SFN exerts distinct molecular effects in different types of cancer. Su et al. (2018) and Kim et al. (2015) explored the biochemical pathways of SFN in ovarian, colon, and gastric cancer cell lines (e.g., HCT116 human colon cancer cells and AGS human gastric cancer cells), demonstrating that SFN inhibits the expression of hypoxia-inducible factor-1α (HIF-1α) and vascular endothelial growth factor (VEGF). Wang et al. (2016) reported that in gastric carcinoma, SFN modulates interleukin-6 (IL-6) signalling by activating microRNA-124 (miR-124), which targets the 3′-untranslated regions (UTR) of IL-6 (Wang et al. 2016).

In bladder cancer, SFN treatment has been shown to significantly induces glutathione S-transferase and NAD(P)H:quinone oxidoreductase 1 (Leone et al. 2017). Additionally, in the Aβ-peptide-stimulated human monocytic leukaemia cell line (THP-1), SFN suppresses IL-1β expression and inhibits microRNA-146a production in microglia-like cells (An et al. 2016). Scientists have identified different effects of sulforaphane on different types of cancer by inhibiting or promoting specific genes. Increased expression levels of Bax, P53, and P27 in ovarian cancer (Kan et al. 2018) and P21 in oral cancer cells (Adtani et al. 2025) prompt cell cycle arrest in the S and G2/M phases and reduce proliferation in cervical and prostate cancer cells (Cheng et al. 2016; Rutz et al. 2020). Hać et al. (2020) demonstrated that SFN (in studies of prostate cancer cells) caused more pronounced double-stranded DNA breaks in cancer cells than in normal cells. This is likely because DNA repair is more efficient in normal cells than in cancer cells. This observation suggests that SFN selectively inhibits cancer cell proliferation in vitro. Livingstone (2022) highlighted that SFN was detected at significantly higher levels in the prostate of men receiving glucoraphanin supplements.

Moreover, multiple in vitro studies have indicated that sulforaphane exerts strong antiproliferative and pro-apoptotic effects in prostate cancer models. In hormone-dependent and hormone-independent prostate cancer cell lines (LNCaP and PC-3), SFN treatment (5 and 15 μM) resulted in a reduction in cell proliferation and modulation of epigenetic regulators, including histone deacetylases (HDACs) and proteins such as GAPDH, MAP1LC3B2, and H2AFY, highlighting its role in autophagy and transcriptional regulation (Beaver et al. 2017). Comparable cytotoxic effects were observed in DU-145 cells treated with broccoli-derived SFN-rich extracts, where approximately 50% of cancer cell death was achieved after 24–48 h of exposure (Türker et al. 2019). Additionally, SFN-induced autophagy has been reported in PC3 cells, suggesting a cytoprotective–cytotoxic dual role depending on the cellular context (Borkowska et al. 2024). The cytoprotective role of SFN occurs through the regulation of redox metabolism, inhibition of histone deacetylases (HDACs) and topoisomerases, modulation of G1 to S-phase progression, and induction of G2/M phase arrest. Additionally, its impact on both phase I and phase II detoxification enzymes plays a critical role in DNA replication and cellular homeostasis (Gao et al. 2021; Mangla et al. 2021; Feng et al. 2023) (Fig. 3.1).

Extensive evidence supports the anti-cancer potential of SFN in various breast-cancer subtypes. In triple-negative breast cancer (TNBC) cell lines (SUM149, SUM159, MDA-MB-231), SFN at micromolar concentrations (2.5–40 μM) significantly inhibited NF-κB signalling, reduced cyclin D1 expression, and suppressed pro-inflammatory cytokines IL-6 and IL-8, leading to apoptosis and impaired migration (Burnett et al. 2017; Bagheri et al. 2020). In oestrogen receptor–positive models (MCF-7 and T47D), SFN downregulated ER-α, EGFR, and HER2 expression and promoted apoptotic cell death after 72 h of treatment (Pledgie-Tracy et al. 2007). Notably, R-sulforaphane (R-SFN) exhibited enhanced induction of Nrf2, NQO1, and GSTP, underscoring the stereoselective differences in chemopreventive efficacy (Licznerska et al. 2021). In lung cancer models, including non-small cell lung cancer (NSCLC) and small cell lung cancer (SCLC), SFN consistently inhibits proliferation, induces apoptosis, and targets cancer stem cell-associated pathways. In A549 and H460 cells, SFN suppressed β-catenin signalling, reduced the expression of stemness markers (Oct4, Sox2, and Nanog), and impaired self-renewal capacity (Xie et al. 2016; Wang et al. 2021). Additional studies have demonstrated SFN-mediated modulation of the GSK3β/β-catenin axis and miR-616-5p expression, supporting its role as a potential adjuvant chemotherapeutic agent (Wang et al. 2017). In SCLC models, SFN induced ferroptosis via downregulation of the cystine/glutamate antiporter xCT (SLC7A11), highlighting a novel mechanism of cell death (Iida et al. 2021). SFN exhibits robust anti-cancer activity in colorectal cancer (CRC) cell lines, primarily by inducing apoptosis and ferroptosis and inhibiting PI3K/Akt/mTOR signalling. In HCT116 and SW620 cells, SFN reduced cell viability and promoted apoptosis by downregulating Bcl-2 and activating Bax (Jiang et al. 2022; Zhu et al. 2022). Recent evidence indicates that SFN induces ferroptosis in HCT116 cells by activating the SIRT3/AMPK/mTOR signalling pathway, further expanding its potential in preclinical studies (Hu et al. 2024). Similar antiproliferative effects were observed in HT-29 cells, with a marked reduction in the number of viable cells following SFN exposure (Đurić et al. 2023). In gastric carcinoma models, SFN demonstrated dose-dependent inhibition of proliferation, migration, and survival pathways. Treatment of MGC803, AGS, and HGC27 cells resulted in the suppression of PI3K/Akt signalling, downregulation of Bcl-2, and induction of Bax- and p53-mediated apoptosis (Dong et al. 2018; Zhang and Tu 2018). Furthermore, SFN modulates epigenetic regulators, such as SMYD3, with overexpression of this histone methyltransferase partially abrogating SFN's anti-cancer effects of SFN, indicating a key epigenetic mechanism (Dong et al. 2018).

In hepatocellular carcinoma (HCC) cell lines (HepG2 and HHL5), SFN reduced cell viability and induced DNA strand breaks in association with Nrf2 activation and glutathione modulation (Liu et al. 2019). Additional studies have demonstrated the inhibition of colony formation and migration, further supporting the role of SFN and its metabolites as hepatoprotective and chemopreventive agents (Liu et al. 2018). In bladder cancer T24 cells, SFN inhibited cell proliferation and growth while modulating Nrf2-dependent detoxification pathways, including GSH, UGT, and COX-2 expression (He et al. 2021). Anti-cancer activity has also been reported in glioblastoma models, where SFN enhances sensitivity to chemoradiotherapy through thioredoxin reductase 1 (TrxR1) inhibition (Ge et al. 2025). In haematological malignancies, such as acute myeloid leukaemia and monoblastic leukaemia, SFN induces dose-dependent cytotoxicity and apoptosis (Koolivand et al. 2018; Akiyoshi et al. 2019).

The anti-cancer efficacy of sulforaphane is fundamentally driven by its ability to disrupt intracellular redox homeostasis through the targeted inhibition of antioxidant defence systems. The primary molecular target in this process is thioredoxin reductase 1 (TrxR1), a selenocysteine-containing flavoenzyme that maintains redox balance by recycling (and reducing) oxidised thioredoxin (Ge et al. 2025). Sulforaphane covalently conjugates with specific residues (Cys59, Cys64, Cys497, and Sec498) within the active centre of TrxR1, effectively suppressing its enzymatic activity and triggering massive accumulation of reactive oxygen species (ROS) in tumour cells. The induction of oxidative stress is recognised as a key mechanism underlying the tumour-suppressive effects of SFN (Ge et al. 2025).

The induction of regulated cell death pathways represents a multifaceted aspect of sulforaphane's pharmacological profile. Beyond the intrinsic mitochondrial pathway, sulforaphane activates the extrinsic Fas/FasL apoptotic pathway by upregulating the expression of death receptor Fas (Joshi et al. 2025). This upregulation is modulated directly and indirectly through ROS production and stabilisation of transcription factors, such as p53 and NF-κB (Asif Ali et al. 2023; Joshi et al. 2025). Within the mitochondria, sulforaphane alters the Bcl-2/Bax ratio, increasing the expression of pro-apoptotic Bax while concomitantly downregulating anti-apoptotic Bcl-2, which facilitates the release of cytochrome c and activates the caspase cascade, specifically caspases 3, 8, and 9 (Alshammari et al. 2025; Joshi et al. 2025). Emerging evidence highlights the capacity of sulforaphane to induce ferroptosis, an iron-dependent form of cell death characterised by the lethal accumulation of lipid peroxides (Zhao et al. 2025a). By depleting intracellular glutathione pools and inhibiting glutathione peroxidase 4 (GPX4) activity, sulforaphane triggers irreversible oxidative damage to the membrane lipids. This coordinated attack on cellular antioxidant mechanisms and iron regulation is particularly effective against treatment-resistant tumours and cancer stem cells (CSCs), whose elevated basal state of oxidative stress renders them uniquely susceptible to ferroptotic induction (Zhao et al. 2025a).

Epigenetic reprogramming is a cornerstone of the long-term chemopreventive and therapeutic activities of sulforaphane. SFN acts as a natural histone deacetylase (HDAC) inhibitor and is established as a broad-spectrum epigenetic modulator that reactivates silenced tumour suppressor genes (Asif Ali et al. 2023; Zhao et al. 2025a). This HDAC inhibitory activity is a key mechanism underlying SFN's anti-tumour effects of SFN in urologic and other cancers, observed in both in vitro and in vivo models without significant toxicity (Zhao et al. 2025a). Through global demethylation and modulation of microRNA expression, epigenetic interventions engage in extensive molecular crosstalk that can reverse abnormal gene transcription profiles in cancer (Asif Ali et al. 2023; Zhao et al. 2025a). Sulforaphane also inhibits DNA methyltransferases, such as DNMT1 and DNMT3B, which enhances apoptosis and cell cycle arrest (Asif Ali et al. 2023). Recent studies have identified protein arginine methyltransferase 5 (PRMT5) as an epigenetic target. By disrupting the PRMT5/MEP50 complex, sulforaphane reactivates tumour suppressor expression and suppresses cancer cell proliferation and invasion in mesothelioma models (Zhao et al. 2025a). The regulation of non-coding RNAs, including microRNAs (such as miR-199a-5p) and long non-coding RNAs, allows sulforaphane to fine-tune diverse cellular functions, such as proliferation, differentiation, and epithelial-mesenchymal transition (Asif Ali et al. 2023; Zarzycka et al. 2026).

Disruption of pro-survival signalling cascades provides another critical layer of anti-cancer activity. Sulforaphane effectively suppresses the PI3K/Akt/mTOR pathway, which is frequently dysregulated in various cancers, including bladder and breast cancers (Joshi et al. 2025; Zarzycka et al. 2026). By inhibiting Akt phosphorylation and abolishing PI3K activity, sulforaphane disrupts the signalling events that drive uncontrolled tumour growth and metabolism (Joshi et al. 2025). In bladder cancer models, the synergy between sulforaphane and chloroquine regulates both the AKT/GSK-3β and mTOR/ULK pathways, leading to effective autophagy inhibition and reduced migration of malignant cells (Zarzycka et al. 2026). Additionally, sulforaphane targets the JAK-STAT3 signalling pathway, preventing the phosphorylation and nuclear translocation of STAT3 (Joshi et al. 2025). This inhibition not only promotes programmed cell death by suppressing anti-apoptotic signals but also impacts the tumour microenvironment by downregulating PD-L1 expression through the covalent modification of STAT1 cysteines, which blocks interferon-gamma-induced immune evasion (IFN-γ) (Zhao et al. 2025a).

The ability of sulforaphane to remodel the tumour immune microenvironment is vital for its clinical success. In glioblastoma, sulforaphane-induced ROS accumulation facilitates the polarisation of tumour-associated macrophages (TAMs) from a tumour-supportive M2-like phenotype to a tumour-suppressive M1-like phenotype. This reprogramming enhances the phagocytic capability of macrophages and increases the secretion of pro-inflammatory cytokines, such as TNF-α and iNOS, thereby sensitising the tumour to chemoradiotherapy (Ge et al. 2025). Furthermore, sulforaphane suppresses the accumulation and immunosuppressive functions of myeloid-derived suppressor cells (MDSCs) by activating the Nrf2 pathway and reducing prostaglandin E2 (PGE2) secretion (Zhao et al. 2025a). In immunocompetent models, sulforaphane recalibrates adaptive immunity by increasing the proportion of pro-inflammatory Th17 cells and decreasing the Treg/Th17 ratio, primarily by inhibiting macrophage migration inhibitory factor (MIF). These systemic immunomodulatory effects underscore the potential of SFN to alleviate immunosuppression and restore effective anti-tumour immunity. The transcriptional program leading to cancer formation is effectively silenced by SFN, which supports the phosphorylation and degradation of β-catenin, suppressing subsequent steps of the process, such as c-Myc and cyclin D1 (Zhao et al. 2025a). In lung cancer models, sulforaphane inhibits the acquisition of smoke-induced CSC-like properties via the IL-6/ΔNp63α/Notch axis (Xie et al. 2019). Similarly, it targets the Sonic hedgehog signalling pathway (inhibition) to regulate pancreatic CSC self-renewal (Rodova et al. 2012). The compound's interaction with ICAM-1 and N-cadherin further highlights its role in inhibiting epithelial-mesenchymal transition (EMT), thereby reducing the metastatic potential and invasiveness of bladder cancer cells (Zarzycka et al. 2026).

Landmark studies have shown that sulforaphane supplementation enhances the detoxification of airborne pollutants and modulates gene expression in men with recurrent prostate cancer (Asif Ali et al. 2023; Zhao et al. 2025a). In breast cancer, pilot studies have confirmed that sulforaphane metabolites are delivered to human breast tissue, where they induce promoter hypermethylation of genes involved in Wnt signalling and inflammation (Zhao et al. 2025a).

Despite its low-toxicity profile, sulforaphane exhibits hormetic behaviour, characterised by a biphasic dose-response, where low doses stimulate beneficial antioxidant effects through Nrf2, while high doses lead to the inhibitory effects required for direct tumour cell killing (Asif Ali et al. 2023; Zhao et al. 2025a). This dual nature necessitates precise dosing and scheduling, especially when used in combination with chemotherapy, to maximise the therapeutic effect and avoid the potential protection of established tumour cells by persistent Nrf2 activation (Zhao et al. 2025a).

3.3 SFN Activity in the Cognitive Disorders

Clinical trials in a dish have shown that sulforaphane is frequently investigated as a potential inhibitor of progressive cognitive disorders (see Table 4.1). In Alzheimer's disease, which is characterised by the accumulation of β-amyloid in neuronal cell bodies, neuronal atrophy, elevated ceroid-lipofuscin concentrations, and neurofibrillary tangle formation, SFN has been found to decrease β-amyloid levels in vitro and reduce cholinergic neuron loss in mouse models, leading to neurobehavioural improvements (Zhang et al. 2014; Hou et al. 2018; Chegeni et al. 2019) (Fig. 3.2).

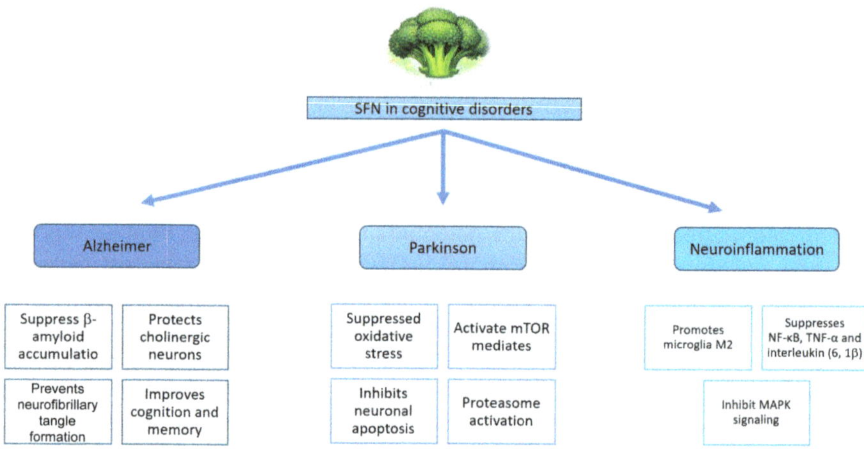

Fig. 3.2 Sulforaphane's activity in the cognitive disorders

In Parkinson's disease, SFN has been shown to mitigate oxidative stress and inhibits neuronal apoptosis in in vitro models by modulating key signalling pathways, including p70S6K (a serine/threonine kinase downstream of PIP3 and phosphoinositide-dependent kinase-1 in the PI3 kinase pathway) and 4E-BP1 (4E-binding protein 1), both of which are mTOR-mediated (Xu et al. 2014; Zhou et al. 2016). Additionally, SFN exerts neuroprotective effects by activating the proteasome, thereby contributing to the maintenance of proteostasis (Santín-Márquez et al. 2019).

Some researchers have proposed that SFN is a natural compound capable of mitigating age-related neuroinflammation. The potentially neuroprotective role of SFN in macrophages residing in the central nervous system (CNS) has been demonstrated in microglial studies. Microglia originate from mesodermal/mesenchymal precursors and migrate into the CNS, where they differentiate into specialised immune cells within the brain microenvironment. They exhibit dynamic activity, interacting closely with both neuronal and non-neuronal cells, and continuously extend and retract their processes to monitor the brain parenchyma for potential threats. As key regulators of synaptic activity and mediators of phagocytosis during neuronal development, microglia play a crucial role in maintaining brain homeostasis and significantly influence synaptic plasticity and neurogenesis (Casano and Peri 2015; Pan et al. 2022; Singh et al. 2022). However, chronic overactivation of microglia leads to the excessive release of neurotoxic mediators, including pro-inflammatory cytokines, reactive oxygen species (ROS), and reactive nitrogen species (RNS), contributing to neurodegenerative pathologies (Park et al. 2015). Persistent microglial activation is strongly associated with neurodegenerative diseases (Von Bernhardi et al. 2015). It has been demonstrated that SFN induces widespread transcriptional changes in genes associated with detoxification and antioxidant responses in microglial cells (Konwiński et al. 2004; Eren et al. 2018). Furthermore, Zhang et al. (2023) reported that SFN reduces the expression of the

MAPK/NF-κB signalling pathway (in Aβ25–35-activated BV-2 cells), which results in the reversal of M1 microglial polarisation and mitigation of neuroinflammation-induced programmed necrosis of neural stem cells (Zhang et al. 2023). In the nervous system, sulforaphane mitigates peripheral neuropathy and cognitive impairment by activating Nrf2-regulated antioxidant defences and reducing the expression of inflammatory enzymes, such as cyclooxygenase-2 (COX-2) and inducible nitric oxide synthase (iNOS) (Mohamadi et al. 2023).

The blood-brain barrier is an important physicochemical mechanism that protects the brain against toxins from cells and the environment, as well as other harmful compounds. Fahey et al. (2025) pointed that SFN has the impact to protect BBB integrity. As BBB integrity is rapidly disrupted in the absence of Nrf2 transcriptional activity (Song et al. 2020; Pan et al. 2022; Cazalla et al. 2024), the likely effect of sulforaphane is through activation of the Keap1/NRF2 pathway. Sulforaphane functions as a potent electrophile that modifies cysteine residues on Kelch-like ECH-associated protein 1 (Keap1), thereby releasing Nrf2 for nuclear translocation and subsequent binding to the antioxidant response element (ARE) (Zheng et al. 2022; Bessetti and Litwa 2025). This transcriptional activation induces a robust phase II detoxification response, upregulating essential enzymes such as NAD(P)H quinone dehydrogenase 1 (NQO1), haeme oxygenase 1 (HO-1), and glutamate-cysteine ligase, which are critical for maintaining redox homeostasis in the highly metabolic environment of the brain (Hei et al. 2022; Yunusoğlu et al. 2025). Furthermore, SFN-filled extracellular vesicles stimulated the proliferation of PC-12 neuronal cells, suggesting a regulatory effect on neuronal gene expression and potential relevance to neurodevelopmental or neurodegenerative diseases (Shahlaei et al. 2022). Parfenova et al. (2020) demonstrated that SFN is a brain-permeable compound that increases cerebral blood flow through a non-genomic mechanism involving cystathionine γ-lyase/cystathionine β-synthase (CSE/cystathionine β-synthase-catalysed, respectively) H_2S production in neurovascular cells, which subsequently activates K_ATP and BK channels in arteriolar smooth muscle.

The neuroprotective capacity of sulforaphane within the central nervous system (CNS) is primarily defined by its ability to cross the blood-brain barrier and accumulate in various brain regions (e.g. striatum), where it reaches peak concentrations relatively quickly following administration (Zheng et al. 2022; Yunusoğlu et al. 2025). This pharmacokinetic efficiency allows the compound to directly interact with neuronal and glial molecular pathways to mitigate the pathological hallmarks of cognitive and neuropsychiatric disorders, such as mitochondrial dysfunction, oxidative stress, and neuroinflammation (Chen et al. 2025; Saito et al. 2025; Yunusoğlu et al. 2025).

In the context of Alzheimer's disease and vascular cognitive impairment, sulforaphane has demonstrated a significant ability to suppress the accumulation of amyloid-beta (Aβ) and hyperphosphorylation of toxic tau proteins (Bessetti and Litwa 2025; Yunusoğlu et al. 2025). However, Leng et al. (2023) in their investigation concluded that "*cortical neuroinflammation coincides with structural and functional network disruption* (is) *independent of Aβ and cortical atrophy. These findings link brain connectivity changes and pathological processes in Alzheimer's disease and suggest a pathway from neuroinflammation to systemic brain dysfunction*."

Preclinical research using transgenic mouse models of Alzheimer's disease has indicated that sulforaphane reduces amyloid plaque burden and improves cognitive performance, specifically in spatial learning and short-term recognition memory (Hou et al. 2018; Yunusoğlu et al. 2025). These effects are often associated with the downregulation of histone deacetylases 1 and 3 (HDAC) and the enhancement of the p75 neurotrophin receptor (Yunusoğlu et al. 2025). Moreover, recent studies on lipopolysaccharide (LPS)-induced recognition memory impairment suggest that sulforaphane exerts its preventive effects by regulating the AMPK/caspase-3 signalling pathway (Alzahrani et al. 2024). In these models, sulforaphane pre-treatment significantly improved performance in novel object recognition and temporal order recognition (NOR and TOR, respectively) tests, probably by increasing phosphorylated AMPK levels and decreasing active caspase-3, thereby preventing neuronal apoptosis and microglial hyperactivity in the prefrontal cortex (Alzahrani et al. 2024). By inhibiting acetylcholinesterase (AChE) overactivation, sulforaphane further supports cholinergic function, which is often compromised in neurodegenerative diseases (Alzahrani et al. 2024; Yunusoğlu et al. 2025).

The immunomodulatory activity of sulforaphane is equally vital for cognitive health because chronic neuroinflammation is a major driver of synaptic loss and cognitive decline. Sulforaphane suppresses the nuclear factor kappa-B (NF-κB) pro-inflammatory cascade, leading to a marked reduction in the production of cytokines such as tumour necrosis factor-alpha (TNF-α), interleukin-1beta (IL-1β), and interleukin-6 (IL-6) (Zheng et al. 2022; Bessetti and Litwa 2025). This suppression is partly achieved through the regulation of mitogen-activated protein kinases (MAPKs), including p38, c-Jun N-terminal (JNK), and extracellular signal-regulated (ERK) kinases (Zheng et al. 2022). Crucially, sulforaphane promotes the polarisation of microglia from a pro-inflammatory M1 phenotype to an anti-inflammatory M2 phenotype, which facilitates tissue repair and removal of cellular debris (Zheng et al. 2022; Bessetti and Litwa 2025). In depression models, the transition of microglia to a neuroprotective state is accompanied by a reduction in methyl-CpG binding protein 2 (MECP2) levels and an increase in Nrf2-mediated brain-derived neurotrophic factor (BDNF) transcription, which preserves synaptic density and reduces depressive-like behaviours (Zheng et al. 2022; Bessetti and Litwa 2025).

Schizophrenia and related psychotic disorders represent another significant area of therapeutic potential for sulforaphane. Cognitive deficits in schizophrenia, such as impairments in working memory and verbal learning, have been linked to oxidative stress and epigenetic silencing of markers such as glutamic acid decarboxylase 67 (GAD67) (Hei et al. 2022; Saito et al. 2025). Clinical trials in patients with first-episode or early stage schizophrenia have shown that while sulforaphane may not significantly alter core psychiatric symptoms as measured by The Positive and Negative Syndrome Scale (PANSS), it produces statistically significant improvements in selective cognitive domains, including spatial working memory and verbal learning (Hei et al. 2022; Chen et al. 2025). These clinical improvements are supported by preclinical evidence showing that sulforaphane can rescue defects in sensorimotor gating and cognitive function in phencyclidine (PCP)-induced models by protecting parvalbumin-positive interneurons from oxidative damage (Hei et al. 2022; Bessetti and Litwa 2025). The ability of sulforaphane to function as an HDAC inhibitor provides an additional epigenetic mechanism for restoring GABAergic function and synaptic plasticity in the prefrontal cortex (Hei et al. 2022; Saito et al. 2025).

In neurodevelopmental disorders, most notably in autism spectrum disorder (ASD), sulforaphane has gained prominence owing to positive results in several clinical trials. Initial studies reported that sulforaphane-rich broccoli sprout extract significantly improved social responsiveness and aberrant behaviours in young men with moderate-to-severe ASD. These behavioural improvements were correlated with the normalisation of redox metabolism markers and the reduction of inflammatory cytokines such as IL-6 and TNF-α (Clark et al. 2010). Furthermore, sulforaphane has been shown to improve mitochondrial parameters, such as ATP-linked respiration, which are often dysfunctional in individuals with ASD (Zheng et al. 2022; Saito et al. 2025). While subsequent trials have shown mixed results regarding the magnitude of clinical improvement, clinician-rated assessments have indicated some gains in social interaction and communication barriers (Bessetti and Litwa 2025; Saito et al. 2025).

The efficacy of sulforaphane in various conditions often follows a hormetic dose-response relationship, in which low concentrations consistently elicit beneficial antioxidant and cytoprotective responses. In neuronal cell cultures, doses as low as 0.1 μM have been shown to maximise cell viability and Nrf2 nuclear translocation, whereas higher concentrations may be detrimental (Bessetti and Litwa 2025). This hormetic effect highlights the importance of precise dosing in clinical applications, especially as neurones appear more susceptible to sulforaphane's activity than other cell types (e.g. cancer cells) (Bessetti and Litwa 2025). Despite its rapid metabolism and relatively short half-life (1.9 ± 0.4 h), the use of stabilised precursors, such as glucoraphanin, especially when combined with active myrosinase, has improved the bioavailability and clinical feasibility of sulforaphane as a long-term intervention (Atwell et al. 2015; Hei et al. 2022; Saito et al. 2025).

In addition to its role in chronic neurodegeneration and developmental disorders, sulforaphane protects against acute brain injuries. In models of traumatic brain injury (TBI), sulforaphane administration reduces cerebral oedema by upregulating aquaporin 4 (AQP4) expression and decreases neuronal death and tissue damage through Nrf2-mediated antioxidant enzyme induction (Yunusoğlu et al. 2025). Similar protective effects have been observed in epilepsy models, where sulforaphane reduces reactive oxygen species (ROS) generation during seizure activity and restores glutathione levels, thereby preventing seizure-induced hippocampal damage and cognitive deficits (Bessetti and Litwa 2025; Yunusoğlu et al. 2025).

3.4 Cardiovascular Protection and Metabolic Disorders

Sulforaphane exhibits significant potential for mitigating oxidative stress, as evidenced by its ability to modulate the expression of antioxidant enzymes, such as catalase, manganese superoxide dismutase (Mn-SOD), and haeme oxygenase-1 (HO-1) (Angeloni et al. 2009; Guerrero-Beltrán et al. 2012). This property has been widely explored in studies on cardiovascular diseases, including hypertension, atherosclerosis, ischaemia-reperfusion injury, peripheral artery diseases, and congestive heart failure. Studies have demonstrated that SFN can open mitochondrial ATP-sensitive potassium (K_ATP) channels and activate antioxidant enzymes, thereby exerting cardioprotective effects, particularly in post-ischaemic conditions (Mangla et al. 2021) (Fig. 3.3).

The cardiovascular and endothelial protective effects of SFN have been documented in cellular models relevant to pregnancy-associated vascular dysfunction. In human umbilical vein endothelial cells exposed to TNF-α, SFN treatment reduced the secretion of endothelin-1 and the expression of adhesion molecules, including VCAM-1, ICAM-1, and E-selectin, while preserving endothelial barrier integrity. Notably, these protective effects extend beyond canonical Nrf2 signalling, suggesting the involvement of additional cytoprotective mechanisms (Cox et al. 2019).

Bai et al. (2013) reported that SFN administration reduced myocardial infarct size, decreased cardiomyocyte apoptosis, cytochrome c release, and procaspase-3 activity, and improved post-ischaemic ventricular function. Additionally, SFN-mediated cardioprotection has been linked to the transient activation of Nrf2, leading to an increased expression of phase I detoxification enzymes at the end of reperfusion (Silva-Palacios et al. 2019). Moreover, SFN-induced regulation of the AKT (activation), GSK-3β (inhibition), and Fyn (accumulation) pathways has been shown to protect against angiotensin II-induced cardiac dysfunction (Xin et al. 2018).

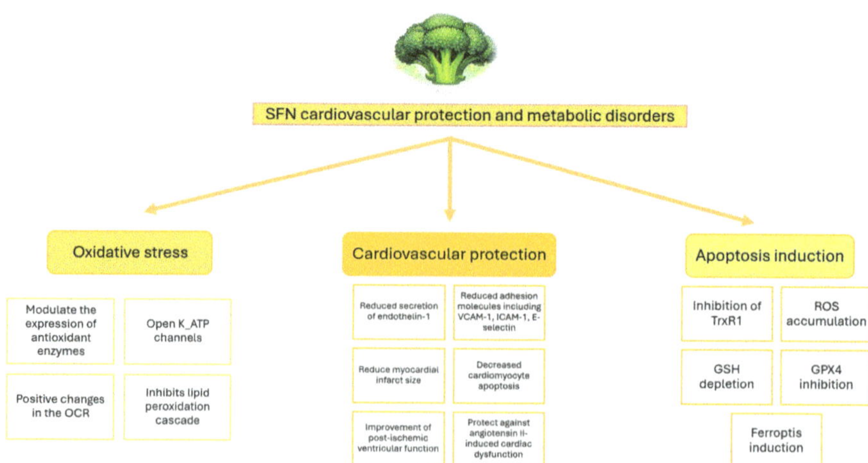

Fig. 3.3 Sulforaphane's in metabolic disorders

The cardioprotective effects of sulforaphane appear to be largely mediated by the modulation of mitochondrial bioenergetics and suppression of oxidative stress arising from metabolic dysregulation (Mthembu et al. 2025; Wang et al. 2025). Initial evidence for these effects has been established at the cellular level using cultured H9c2 cardiomyoblasts, where pre-treatment with sulforaphane (10 μM) enhanced metabolic activity under lipidemic stress conditions induced by palmitic acid exposure. This metabolic improvement was observed as a marked increase in ATP in the cytoplasm, improvements in mitochondrial respiration status, and positive changes in the oxygen consumption rate (OCR) and extracellular acidification rate (ECAR), which possibly occurs due to the reduction of proton (H^+) leakage across the mitochondrial membrane. Therefore, SFN mitigates energetic imbalances and membrane gradient instabilities that typically predispose individuals to cardiovascular complications (Mthembu et al. 2025). At the molecular level, these functional improvements are accompanied by upregulation of the mRNA expression of key mitochondrial biogenesis markers, including nuclear respiratory factor 1 (Nrf1), peroxisome proliferator-activated receptor gamma coactivator 1-alpha (Ppargc1α), and ubiquinol-cytochrome c reductase complex assembly factor 1 (Uqcc1) (Mthembu et al. 2025). Administration of sulforaphane at 10 mg/kg softens the characteristic features of diabetic cardiomyopathy (DCM) and metabolic syndrome (MetS), including hyperglycaemia, dyslipidaemia, and insulin resistance, as measured by the HOMA-IR index (Masoumvand et al. 2025; Wang et al. 2025). In the myocardium, sulforaphane inhibits cardiolipotoxicity, probably by influencing the lipid peroxidation cascade, which is reflected by reduced levels of malondialdehyde (MDA) and superoxide anions (Zhang et al. 2022; Wang et al. 2025). SFN modulates fatty acid transporters and metabolic regulators, downregulating the expression of CD36 (Cluster of Differentiation 36), cardiac-type fatty acid binding protein (H-FABP), carnitine palmitoyltransferase 1B (CPT1B), and peroxisome proliferator-activated receptor alpha (PPARα) in the myocardium. By inhibiting these key mediators, sulforaphane limits excessive and uneven uptake and oxidation of fatty acids, which are key factors in mitochondrial and diastolic dysfunction in the diabetic heart (Wang et al. 2025). Importantly, the effect on lipid metabolism appears to be more pronounced than that on glucose metabolism, as the expression of glucose transporters, such as GLUT4 and mitochondrial pyruvate transporters (MPC1/2), often remains unchanged after treatment (in chronic experimental models), suggesting a substrate-specific remodelling of cardiac metabolism rather than a general metabolic correction (Wang et al. 2025).

SFN-enhanced antioxidant capacity in cardiac tissue plays a key role in neutralising reactive oxygen species (ROS) and preventing oxidative damage, leading to apoptosis and cell necrosis (Mthembu et al. 2025). Experimental evidence suggests that activation of the Nrf2/HO-1 axis by sulforaphane not only protects against hyperglycaemia-induced damage but also maintains mitochondrial membrane potential and mass, preserving the structural integrity of the myocardium under conditions of metabolic stress (Mohamadi et al. 2023; Mthembu et al. 2025). Clinical studies have confirmed these findings, indicating that individuals at risk of cardiovascular disease who consume sulforaphane-rich dietary sources experience improved oxidative status and normalised cholesterol levels (Alves et al. 2025).

In the context of vascular health and atherosclerosis, sulforaphane demonstrates significant anti-inflammatory and anti-atherogenic properties by inhibiting the nuclear factor kappa-light-chain-enhancer of activated B cells (NF-κB) pathway (Alves et al. 2025). By suppressing NF-κB DNA-binding activity, sulforaphane reduces the expression of pro-inflammatory cytokines and adhesion molecules, such as intercellular adhesion molecule 1 (ICAM-1) and vascular cell adhesion molecule 1 (VCAM-1) in endothelial cells (Alves et al. 2025; Zhao et al. 2025b). This prevents monocyte adhesion to the endothelium and limits the formation of atherosclerotic lesions (Alves et al. 2025). In addition, sulforaphane counteracts the unfavourable effects of advanced glycation end products on endothelial cells and reduces the activation of NADPH oxidase (Masoumvand et al. 2025; Zhao et al. 2025b). SFN also exhibits antithrombotic effects (preventing platelet aggregation and thrombus formation) by inhibiting the PI3K/Akt pathway in platelets (Jayakumar et al. 2013; Alves et al. 2025). These vascular protective effects are complemented by antihypertensive effects, as sulforaphane has been shown to reduce mean arterial pressure and prevent vascular wall thickening in hypertensive models (Masoumvand et al. 2025). One identified mechanism for blood pressure regulation involves the modulation of arachidonic acid metabolism in the kidney, specifically the inhibition of 20-hydroxyeicosatetraenoic acid (20-HETE) production, which facilitates vasoconstriction and sodium reabsorption (Elbarbry et al. 2014).

The maintenance of metabolic health throughout the lifespan is increasingly recognised as a challenge due to the rising prevalence of obesity and related low-grade chronic inflammation (Alves et al. 2025). Sulforaphane intervention has been shown to mitigate high-fat-diet-induced obesity through multiple mechanisms, including the induction of adipocyte browning and enhancement of mitochondrial biogenesis in skeletal muscle (Alves et al. 2025; Zhao et al. 2025b). The antiobesogenic effects of sulforaphane are mediated by the activation of the AMPK-mTOR-ULK1 signalling pathway, which facilitates lipophagy and the subsequent reduction of visceral adiposity and fat accumulation in the liver (Zhao et al. 2025b). Additionally, sulforaphane suppresses adipogenesis by downregulating the expression of key transcription factors, such as peroxisome proliferator-activated receptor gamma (PPARγ) and enhancer-binding protein alpha (C/EBPα). A significant finding in recent geroscience research is sulforaphane's ability to reverse leptin resistance, a major driver of obesity-associated metabolic dysfunction (Çakır et al. 2022; Alves et al. 2025). By restoring leptin responsiveness, sulforaphane helps regulate food intake and body weight gain, thereby reducing the systemic burden that leads to cardiovascular complications (Shawky and Segar 2018; Alves et al. 2025; Masoumvand et al. 2025). Metabolic disorders are another important area of SFN activity. In vitro studies assessing the antidiabetic potential of sulforaphane-rich broccoli extracts have shown that they inhibit α-amylase activity in a dose-dependent manner, supporting their hypoglycaemic properties (Basha et al. 2017). Complementary human intervention studies have demonstrated that the consumption of broccoli sprout powder (5–10 g/day) significantly reduces serum insulin concentrations in individuals with insulin resistance, indicating improved metabolic control and insulin sensitivity (Bahadoran et al. 2012). SFN can suppress key enzymes involved in gluconeogenesis, suggesting its potential as a therapeutic agent for hyperglycaemia control (Axelsson et al. 2017). Furthermore, SFN has been reported to improve insulin resistance and attenuate inflammation and oxidative damage associated with diabetes (Xue et al. 2008; Bahadoran et al. 2012; Jiang et al. 2014). By reducing glycated haemoglobin (HbA1c) levels and fasting blood glucose, SFN may be particularly beneficial for individuals with obesity and poorly controlled type 2 diabetes (Axelsson et al. 2017).

The therapeutic potential of sulforaphane in managing type 2 diabetes mellitus is characterised by its ability to enhance insulin sensitivity and promote glucose homeostasis (Bernuzzi et al. 2023). Sulforaphane facilitates glucose uptake in skeletal muscle and liver cells by upregulating the insulin receptor substrate 1 (IRS-1)/PI3K/Akt signalling pathway and promoting the translocation of glucose transporter type 4 (GLUT4) to the plasma membrane (Mohamadi et al. 2023; Masoumvand et al. 2025). Beyond its hypoglycaemic effects, sulforaphane provides significant protection against chronic diabetic complications, including nephropathy and neuropathy (Mohamadi et al. 2023). In diabetic nephropathy models, sulforaphane treatment prevents renal fibrosis and basement membrane thickening by inhibiting histone deacetylase 2 (HDAC2) and upregulating bone morphogenetic protein 7 (BMP-7) expression (Bai et al. 2013; Kong et al. 2021; Mohamadi et al. 2023). Activation of the Nrf2/PINK1 pathway by sulforaphane also maintains mitochondrial homeostasis in podocytes, preventing renal cell death and albuminuria.

Hepatic metabolic health is another critical area in which sulforaphane exerts a protective influence, particularly in the management of metabolic dysfunction-associated steatotic liver disease (part of former metabolic dysfunction-associated steatotic liver disease (MASLD)) (Zhao et al. 2025b). Studies have indicated that sulforaphane can reduce hepatic glucose production and improve glucose tolerance by restoring the interactions between the endoplasmic reticulum (ER) and mitochondria at the mitochondria-associated membranes (MAMs) (Mohamadi et al. 2023). Sulforaphane effectively inhibits hepatic steatosis and prevents the development of metabolic dysfunction-associated steatohepatitis (MASH) by alleviating endoplasmic reticulum (ER) stress through modulation of the AMPK/PPARα axis and reducing the expression of fatty acid synthase (FAS) (Zhao et al. 2025b). Additionally, hepatic lipid utilisation and antioxidant potential are supported by the regulation of the fibroblast growth factor 21 (FGF21)/FGFR1 signalling pathway. Furthermore, the systemic suppression of inflammation and metabolic disorders by SFN is mediated by the regulation of the gut-liver axis, characterised by an increase in intestinal tight junction proteins and a decrease in serum lipopolysaccharide (LPS) levels (Xu et al. 2023; Zhao et al. 2025b).

While current human trials have demonstrated that sulforaphane can significantly reduce fasting blood glucose, HbA1c, and oxidised LDL levels, more extensive large-sample randomised controlled trials are necessary to determine the optimal dosing strategies for diverse populations (Alves et al. 2025; Zhao et al. 2025b).

Furthermore, the protective role of sulforaphane in highly energetic tissues, such as the myocardium, emphasises its importance in maintaining systolic function and preventing the structural modification of heart tissue, known as cardiac remodelling (Wang et al. 2025). In diabetic models, this remodelling is often characterised by diffuse fibrosis and diastolic dysfunction, both of which markedly improve following sulforaphane intervention (Fix et al. 2019; Wang et al. 2025). These findings provide solid preclinical evidence that chronic oral administration of sulforaphane is a viable strategy for mitigating the detrimental consequences of lipid overload and metabolic stress on the heart (Wang et al. 2025).

3.5 Anti-inflammatory and Immunomodulatory Effects

SFN's ability to suppress the production of pro-inflammatory interleukins and chemokines, including IL-6, IL-8, and TNF-α, suggests its potential therapeutic application in autoimmune and chronic inflammatory diseases (Costa-Pérez et al. 2024). Specifically, SFN inhibits the proliferation of synovial fibroblasts, downregulates matrix metalloproteinases (MMPs) and cyclooxygenase-2 (COX-2), and reduces prostaglandin E2 (PGE2) production, which are key factors implicated in synovial inflammation and the pathogenesis of rheumatoid arthritis (Choi et al. 2014). Additionally, Davidson et al. (2017) reported high concentrations of isothiocyanates in the synovial fluid of individuals with high broccoli intake, suggesting potential chondroprotective effects in osteoarthritis. These findings highlight SFN as a promising dietary compound for joint health and inflammatory disease management. Mohanty et al. (2018) reported that oral administration of SFN (2–5 mg/kg; rats) reduced levels of pro-inflammatory cytokines, including interleukin-6 (IL-6) and tumour necrosis factor-α (TNF-α), and the number of fibroblasts, while simultaneously increasing antioxidant defences, such as glutathione, catalase, and superoxide dismutase activity, and the synthesis of collagen and mucopolysaccharide. Similar anti-inflammatory effects were observed in LPS-stimulated murine macrophages (RAW264.7), where SFN suppressed NF-κB activity, inhibited inflammasome activation, and reduced IL-1β production, suggesting direct interference with innate immune signalling cascades (Treasure et al. 2024). This anti-inflammatory mechanism of sulforaphane involves the downregulation of lipopolysaccharide (LPS)-induced inflammation (Choi et al. 2014; Eren et al. 2018; Subedi et al. 2019). LPS, an endotoxin located on the outer membrane of gram-negative bacteria, triggers a strong inflammatory cascade upon release into the bloodstream (Gorman and Üçışık 2022). Studies on SFN's activity of SFN against LPS-induced acute lung injury have confirmed its anti-inflammatory effects (Qi et al. 2016).

SFN has also been shown to exert protective effects against inflammation-driven organ injury. In mouse models of acute lung injury and acute respiratory distress syndrome (ARDS), SFN administration (50 mg/kg) significantly decreased lung damage markers, including lactate dehydrogenase, IL-6, TNF-α, and prostaglandin E2, even when administered after disease induction, indicating therapeutic rather than prophylactic efficacy (Qi et al. 2016). These findings align with observations that SFN can downregulate cyclooxygenase-2 and inducible nitric oxide synthase, which are key mediators of inflammatory tissue injury (Ruhee and Suzuki 2020). The immunomodulatory role of SFN has been further confirmed in infectious disease models. In systems based on macrophages infected with *Mycobacterium tuberculosis* and *M. smegmatis*, SFN increased bactericidal activity and reduced intracellular bacterial survival, which was possible due to increased expression of the autophagosome marker LC3 II, indicating that SFN may enhance host defence mechanisms against intracellular pathogens (Zhao et al. 2022; Pötschke et al. 2025).

The molecular structure of sulforaphane, which exerts anti-inflammatory and immunomodulatory effects, is based on its ability to modulate key transcription factors and epigenetic regulators, mainly via the reactive electrophilic isothiocyanate moiety. Central to its mechanism is the inhibition of the nuclear factor kappa-light-chain-enhancer of activated B cells (NF-κB) signalling pathway, which serves as a master regulator of the pro-inflammatory response (Treasure et al. 2023). Pretreatment with sulforaphane or its monovalent glycoconjugates has been shown to significantly attenuated the activation of the p65 NF-κB subunit in human dendritic cells (DCs) exposed to lipopolysaccharide (LPS), thereby reducing the secretion of pro-inflammatory cytokines, such as TNF-α, IL-1β, and IL-6 (Leiva-Castro et al. 2025). This inhibitory effect is achieved by preventing the phosphorylation and subsequent proteasomal degradation of the inhibitor protein IκB, which sequesters NF-κB in the cytoplasm (Treasure et al. 2023). Interestingly, the modulation of these pathways appears to be concentration-dependent and can be further refined with mannose- or fucose (6-deoxy-L-galactose)-functionalised SFN-glycoconjugates, which enhance solubility and targeting toward specific immune receptors, such as C-type lectin receptors (CLRs) on dendritic cells (Leiva-Castro et al. 2025).

Nrf2 activation by SFN not only neutralises reactive oxygen species (ROS) but also creates regulatory crosstalk that suppresses NF-κB-driven inflammation. In murine models of chronic allergic airway disease, L-sulforaphane (LSF) administration has been found to increase Nrf2 foci formation within the bronchial epithelium, which correlates with the reversal of airway remodelling features, such as epithelial thickening and subepithelial collagen deposition (Royce et al. 2022). Furthermore, the activation of Nrf2-mediated antioxidant defences in skeletal muscle and the liver has been shown to ameliorate exercise-induced inflammatory stress, reducing markers of muscle damage, such as creatine kinase and pro-inflammatory cytokines, such as IL-6 (Ruhee and Suzuki 2024).

The immunomodulatory profile of sulforaphane extends to the regulation of dendritic cell maturation and T-lymphocyte polarisation. In inflammatory microenvironments, sulforaphane promotes a more immature or "tolerogenic" state in dendritic cells, characterised by the downregulation of essential costimulatory markers, such as CD80, CD83, and CD86, as well as the antigen-presenting molecule HLA-DR (Fernandez-Prades et al. 2023; Leiva-Castro et al. 2025). This phenotypic shift is critical for establishing immune tolerance, as it reduces the ability of dendritic cells to drive effector T-cell responses. Specifically, sulforaphane suppresses the Th2-mediated proliferative response, which is often hyperactivated in conditions such as asthma and food allergies, leading to reductions in IL-5, IL-9, and IL-13 levels (for the last two, significantly) (Fernandez-Prades et al. 2023). In parallel, sulforaphane intervention encourages the proliferation of regulatory T cells (Tregs) and increases the production of the potent anti-inflammatory cytokine IL-10 (Fernandez-Prades et al. 2023; Leiva-Castro et al. 2025). The induction of a stable regulatory immune pattern, along with an increased Treg/Th2 cell ratio, underscores the potential of sulforaphane in rebalancing the immune system under chronic inflammatory conditions.

In addition to transcription factor modulation, sulforaphane exerts profound immunomodulatory effects by inhibiting inflammasome complexes, particularly the NLRP3 inflammasome. The assembly and activation of the NLRP3 complex are responsible for the maturation and secretion of IL-1β and IL-18, as well as the induction of inflammatory cell death. Sulforaphane inhibits both the priming phase of the inflammasome (via NF-κB suppression) and the subsequent activation step, possibly by inhibiting mitochondrial ROS or directly modifying cysteine residues within the inflammasome components (Treasure et al. 2023; Ruhee and Suzuki 2024). However, research conducted by Greaney et al. (2016) pointed that the inhibition of inflammasomes "*is independent of the transcription factor nuclear factor erythroid-derived 2-like factor 2 (Nrf2) and the antioxidant response-element pathway*" (cyt. Greaney et al. 2016).

Studies on microglial cells and macrophages have indicated that sulforaphane effectively reduces the levels of NLRP3 protein and downstream caspase-1 activity, thereby mitigating the hyperinflammatory states observed in diseases ranging from gout to COVID-19 (Kiser et al. 2021; Treasure et al. 2023; Ruhee and Suzuki 2024). Additionally, sulforaphane interacts with Macrophage Migration Inhibitory Factor (MIF), a pleiotropic cytokine that serves as a key regulator of innate immunity and helps override the immunosuppressive effects of glucocorticoids. By covalently modifying the catalytic proline residue of MIF and inhibiting its tautomerase activity, sulforaphane acts as a potential "*steroid-sparing*" agent, which could enhance the effectiveness of corticosteroid therapies while allowing for reduced dosage and fewer side effects (Treasure et al. 2023).

Epigenetic mechanisms also play a significant role in the anti-inflammatory effects of sulforaphane. It has been identified as a natural histone deacetylase (HDAC) inhibitor that specifically targets class I and class II HDAC enzymes, such as HDAC6 and HDAC8 (Royce et al. 2022; Treasure et al. 2023). Molecular modelling and enzymatic assays suggest that sulforaphane and its metabolites, such as SFN-cysteine and SFN-N-acetylcysteine, interact with the catalytic domain of these enzymes, leading to the hyperacetylation of core histones (H2B in reversal and prevention models, and H3 and H4 in only reversal models) and non-histone proteins such as α-tubulin (Royce et al. 2022). This hyperacetylation promotes a more open chromatin conformation that can facilitate the transcription of cytoprotective genes while simultaneously repressing the expression of pro-inflammatory chemokine genes such as CXCL1 and CXCL5 (Royce et al. 2022; Treasure et al. 2023). In the context of allergic airway disease, this epigenetic reprogramming contributes to the suppression of bronchial mucosal damage and restoration of normal lipid distribution within the epithelium (Royce et al. 2022).

Furthermore, the impact of sulforaphane on cellular survival processes, such as apoptosis and autophagy, provides additional layers to its immunomodulatory capacity. In chronic inflammatory microenvironments, sulforaphane increases the rate of apoptosis in inflammatory dendritic cells and cancer cell lines, such as THP-1, while simultaneously decreasing autophagy (Fernandez-Prades et al. 2023). The reduction of LPS-induced autophagy by sulforaphane may enhance its pro-apoptotic effects,

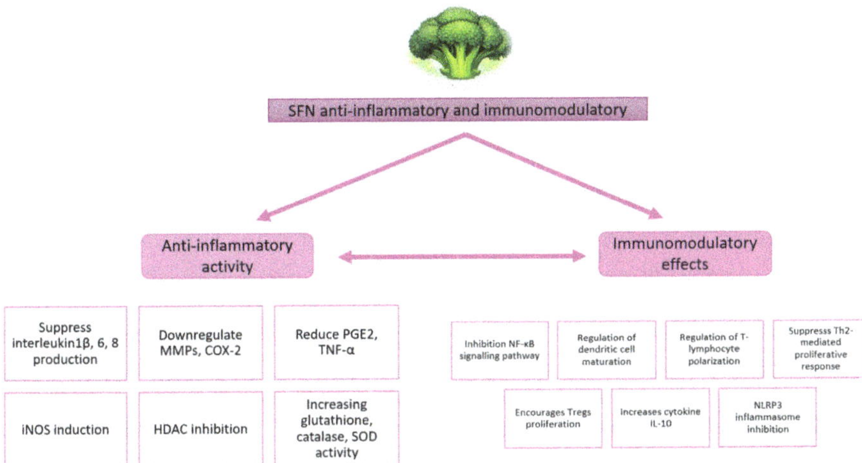

Fig. 3.4 Anti-inflammatory and immunomodulatory activities of sulforaphane

which is proposed as a therapeutic strategy for resolving chronic inflammation and controlling malignant cell proliferation (Fernandez-Prades et al. 2023).

The integrative role of sulforaphane in improving the immune system and healthy aging also has direct implications for cardiovascular and metabolic stability (Alves et al. 2025). Aging is associated with a progressive loss of Nrf2 activity, which increases susceptibility to oxidative stress-induced damage in various tissues. By restoring Nrf2-ARE binding activity, sulforaphane supplementation may delay aging-induced mitochondrial dysfunction and sarcopenia, thereby preserving cardiac and muscular function in late adulthood. Sulforaphane's ability to inhibit Toll-like receptor 4 (TLR4) oligomerisation, and the subsequent activation of the NLRP3 inflammasome further protects the body against the "*cytokine storm*" and chronic inflammatory states that exacerbate metabolic syndrome (Alves et al. 2025; Zhao et al. 2025b) (Fig. 3.4).

3.6 Antimicrobial and Antifungal Properties

Sulforaphane exerts broad-spectrum antimicrobial and antifungal effects against various pathogens, including *Bacillus cereus* and *Staphylococcus aureus* (gram-positive), *Escherichia coli*, *Vibrio cholerae*, and *Pseudomonas aeruginosa* (gram-negative), as well as fungal species such as *Candida albicans* and *Aspergillus flavus* (Devi and Thangam 2012). However, Marshall et al. (2023) reported a significant increase in the growth of enteropathogenic *Escherichia coli* strain EPEC E2348/69 in response to SFN, which increased anaerobic respiration simultaneously with increased anti-inflammatory (for human cells) metabolites. This suggests that its antimicrobial effects may be strain-dependent, warranting further investigation.

The antimicrobial potential of sulforaphane encompasses a broad spectrum of inhibitory actions against diverse bacterial and fungal species, offering a promising phytochemical alternative to conventional antibiotics and fungicides. Research has consistently demonstrated that SFN exerts potent growth-inhibiting properties against various Gram-positive and Gram-negative pathogens, including antibiotic-resistant strains (Pötschke et al. 2025).

Sulforaphane has shown significant effectiveness against prominent gram-negative pathogens, particularly those affecting the gastrointestinal and respiratory tracts. One of the most extensively studied targets is *Helicobacter pylori*, the primary causative agent of gastritis and gastric cancer (Haristoy et al. 2003; Mirmiran et al. 2017; Guo et al. 2022). SFN-enriched extracts have been shown to reduce colonisation and attenuate pathogen-induced gastritis in both murine models and human clinical studies, with minimum inhibitory concentrations (MICs) ranging from 0.06 to 8 μg/mL (Pötschke et al. 2025). Mangla et al. (2021) attributed this effect to SFN's ability to inhibit bacterial enzymes essential for redox balance and metabolism, as well as its disruption of bacterial cell membrane integrity. Furthermore, SFN acts as an irreversible inhibitor of urease, a critical enzyme for *Helicobacter pylori* survival in the acidic gastric environment (Fahey et al. 2013).

In the context of multi-drug resistant (MDR) infections, SFN has demonstrated potent activity against *Pseudomonas aeruginosa*. Recent studies have revealed that SFN effectively reduces bacterial motility and biofilm formation while interfering with the oxidative stress responses of pathogens (Bendary et al. 2024). Suforaphane suppresses the expression of genes involved in systems that regulate the production of virulence factors (quorum sensing), thereby dampening bacterial virulence and reducing immunopathology in infected hosts (Bendary et al. 2024).

SFN inhibits *Shigella dysenteriae* by negatively regulating the expression of Shiga toxin, a primary virulence factor (Nowicki et al. 2021a; He et al. 2024). In vivo studies using the *Galleria mellonella* model demonstrated that SFN treatment significantly improved survival rates in *S. dysenteriae* infections, showing additive effects when combined with synthetic antibiotics such as azithromycin (Krause et al. 2021). The effectiveness of SFN extends to *Escherichia coli* and *Salmonella* species. Although individual isolates may vary in susceptibility, SFN frequently exhibits lower MICs than synthetic positive controls in specific experimental setups. Furthermore, SFN has been identified as a candidate for the inhibition of horizontal gene transfer. Although its effects as an anti-conjugant are considered minor, it has been shown to induce plasmid curing in certain *E. coli* strains (Pötschke et al. 2025).

Gram-positive bacteria, including *Staphylococcus aureus* and *Enterococcus faecalis*, are highly susceptible to the inhibitory effects of sulforaphane (Aires et al. 2009). In studies involving clinical isolates, SFN displayed MIC values ranging from 1 to 4 mg/L against a wide variety of gram-positive bacteria, including *B. subtilis* and *E. faecalis* (Pötschke et al. 2025). It has been observed that SFN can clear both extracellular and intracellular *S. aureus* in macrophages by inhibiting pro-inflammatory signalling pathways such as JNK and p38 MAPK, thereby reducing bacterial survival without inducing cellular apoptosis (Deramaudt et al. 2020).

A significant area of recent application of SFN is in oral health and the prevention of dental carries. *Streptococcus mutans* is recognised as the dominant cariogenic bacterium because of its ability to form biofilms and produce acids

(Welin-Neilands and Svensater 2007; Yu et al. 2025). Sulforaphane exhibits bacteriostatic activity against *S. mutans* with an MIC of 256 μg/mL (rats) and significantly weakens its cariogenic properties, even at sub-inhibitory concentrations. Phenotypic experiments have confirmed that SFN suppresses the synthesis of extracellular polysaccharides (EPS), reduces lactic acid production, and decreases acid tolerance (Yu et al. 2025). Genetic analysis via qRT-PCR revealed that SFN downregulates critical virulence genes, such as gtfB, gtfC, and gtfD (related to EPS synthesis), ldh (acid production), and atpD (acid tolerance), and inhibits the comCDE quorum-sensing system, which is essential for biofilm maturation (with no observed toxicity) (Yu et al. 2025).

Sulforaphane exhibits clear antifungal and fungicidal activity against *Candida albicans* and other non-albicans species, such as *C. glabrata* and *C. krusei*. In vitro assays have shown that SFN typically possesses an MIC of 30 μg/mL against various *C. albicans* strains, including oral isolates from immunocompromised patients. Additionally, SFN targets the transition of yeast to hyphae, a structural change that is essential for fungal colonisation and tissue invasion (Silva et al. 2022). At sub-inhibitory concentrations, SFN significantly impaired biofilm formation and reduced the expression of hyphae-related virulence genes, including ALS1, EFG1, and RAS1. In silico analyses support these findings, predicting that SFN acts through mechanisms such as the inhibition of yeast RNA and alteration of membrane composition. Furthermore, SFN has shown synergistic and additive effects when combined with fluconazole, suggesting its potential as an adjuvant therapy to enhance clinical outcomes and reduce the required dosage of standard antifungals (Silva et al. 2022). The antifungal properties of sulforaphane have gained attention as an alternative to conventional azole therapies, which are facing increasing resistance (Fig. 3.5).

Beyond human health, SFN is a potent biocontrol agent against plant pathogens, offering an eco-friendly alternative to chemical fungicides. SFN-enriched extracts from broccoli have been shown to remarkably inhibit the proliferation of pathogens, such as *Xanthomonas campestris, Fusarium oxysporum, Phytophthora infestans,* and *Gibberella zeae* (He et al. 2024). Molecular docking simulations have shown that SFN and related indole-derived GHPs interact with key detoxifying enzymes in *B. cinerea*, such as eburicol 14-alpha-demethylase (CYP51) and glutathione S-transferase (GST). CYP51 inhibition is particularly important because it disrupts ergosterol biosynthesis, a key component of the fungal cell membrane, leading to cell death (Román et al. 2024). The site and efficacy of action are important when using SFN in fungi. Román et al. (2024) reported that in *Botrytis cinerea* (the causative agent of "gray mold"), sulforaphane and other glucosinolate hydrolysis products were more effective in inhibiting spore germination than mycelial growth. In addition to direct killing, exogenous application of SFN activates the reactive oxygen species (ROS) signalling pathway in plants, priming their internal defence mechanisms against pathogenic infections. Research has also identified the upregulation of defence-related genes, such as NPR1 (Nonexpressor of Pathogenesis-Related Genes 1), PR1 (Pathogenesis-Related Protein 1), and PDF1.2 (Plant Defensin 1.2), in plants treated with SFN-enriched extracts (He et al. 2024).

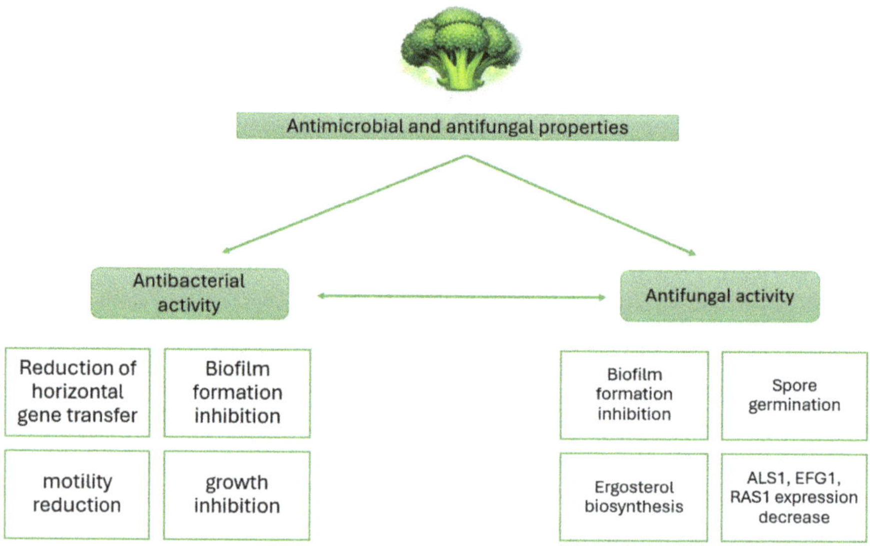

Fig. 3.5 Sulforaphane's antimicrobial and antifungal activities

The antimicrobial activity of sulforaphane is attributed to its highly reactive isothiocyanate group (-NCS), which has a strong affinity for sulfhydryl groups in proteins and enzymes (Andrés et al. 2025). This reactivity allows SFN to disrupt multiple cellular processes across different species.

The primary mechanism of action of SFN is the destruction of pathogenic cell membranes. He et al. (2024) used fluorescence microscopy (with propidium iodide (PI) staining) to show that pathogens treated with SFN exhibit severe membrane permeabilisation. This structural damage leads to the collapse and tortuosity of bacterial bodies, as observed in *Xanthomonas campestris* and *Staphylococcus aureus* (He et al. 2024). In bacteria, SFN induces a "*stringent response*," which is a survival mechanism triggered by environmental stress. This is evidenced by the increased accumulation of (p)ppGpp alarmones following exposure to SFN (Nowicki et al. 2021b; Pötschke et al. 2025). This response involves the modulation of the relA-mediated pathway, which ultimately interferes with normal bacterial metabolism and growth (Nowicki et al. 2021b). Sulforaphane targets specific enzymes that are essential for pathogen survival. This includes the irreversible inhibition of urease in *H. pylori* and targeting of the type III secretion system (TTSS) in *Pseudomonas syringae*, which is necessary for the injection of virulence proteins into host cells (Fahey et al. 2013; He et al. 2024). In fungi, the interaction between CYP51 and GST enzymes provides a clear pathway for mycelial growth suppression (Román et al. 2024). By inducing Phase II detoxification enzymes and modulating pro-inflammatory cytokines, SNF creates an environment that favours bacterial clearance while protecting host tissues from inflammatory damage (Pötschke et al. 2025).

In addition to sulforaphane, broccoli also contains a variety of antimicrobial peptides (AMPs) that contribute to its overall defence profile. These peptides, such as ARFEELNMDLFR (dodecapeptide) and SIWYGPDRP Class I defensins (BraDef), target bacterial membranes and create pores that lead to ion leakage and cell lysis (Pacheco-Cano et al. 2020). These AMPs have demonstrated broad-spectrum activity against *Bacillus cereus, Salmonella, Listeria monocytogenes*, and phytopathogenic fungi such as *Aspergillus niger* and *Colletotrichum gloeosporioides. They* are being explored for applications in active food packaging and as natural preservatives (Nicolas-Espinosa et al. 2022; Andrés et al. 2025).

References

Adtani PN, Al-Bayati SAAF, Elsayed WS (2025) Sulforaphane from *Brassica oleracea* induces apoptosis in oral squamous carcinoma cells via p53 activation and mitochondrial membrane potential dysfunction. Pharmaceuticals 18:393. https://doi.org/10.3390/ph18030393

Aires A, Mota VR, Saavedra MJ, Rosa EAS, Bennett RN (2009) Antimicrobial effects of glucosinolates and their hydrolysis products on human intestinal bacteria. J Appl Microbiol 106:2086–2095

Akiyoshi S, Kikuchi H, Kuribayashi F, Madhyastha H, Minami H (2019) Sulforaphane displays growth inhibition and cytotoxicity in U937 cells. Fundam Toxicol Sci 6:319–325

Alhazmi N, Pai CP, Albaqami A, Wang H, Zhao X, Chen M et al (2020) PML1 is an oncoprotein and direct target of sulforaphane. BBA Mol Cell Res 1867:118707

Alshammari AN, Alfawaz MS, Alenezi YM, Al-Ghafari AB, Al Doghaither HA, Elmorsy EM et al (2025) Nanoliposome-mediated sulforaphane delivery suppresses Ehrlich carcinoma. BMC Cancer 25:1838

Alves I, Araújo EMQ, Dalgaard LT, Singh S, Børsheim E, Carvalho E (2025) Protective effects of sulforaphane against inflammation and oxidative stress. Nutrients 17:428

Alzahrani NA, Bahaidrah KA, Mansouri RA, Aldhahri RS, Abd El-Aziz GS, Alghamdi BS (2024) Prophylactic effects of sulforaphane on LPS-induced memory impairment. Biomedicine 12:1107

An YW, Jhang KA, Woo SY, Kang JL, Chong YH (2016) Sulforaphane exerts anti-inflammatory effects via STAT-1 dephosphorylation and Nrf2/HO-1 activation. Neurobiol Aging 38:1–10

Andrés CMC, Pérez de la Lastra JM, Munguira EB, Juan CA, Pérez-Lebeña E (2025) Health benefits of broccoli: glucosinolates, phenolics and antimicrobial peptides. Molecules 30:2262

Angeloni C, Leoncini E, Malaguti M, Angelini S, Hrelia P, Hrelia S (2009) Modulation of phase II enzymes by sulforaphane: cardioprotective implications. J Agric Food Chem 57:5615–5622

Asif Ali M, Khan N, Kaleem N, Ahmad W, Alharethi SH, Alharbi B et al (2023) Anticancer properties of sulforaphane. Front Oncologia 13:1168321

Atwell LL, Hsu A, Wong CP, Stevens JF, Bella D, Yu TW et al (2015) Absorption and chemopreventive targets of sulforaphane in humans. Mol Nutr Food Res 59:424–433

Axelsson AS, Tubbs E, Mecham B, Chacko S, Nenonen HA, Tang Y et al (2017) Sulforaphane reduces hepatic glucose production in type 2 diabetes. Sci Transl Med 9:4477

Bagheri M, Fazli M, Saeednia S, Kharanagh MG, Ahmadiankia N (2020) Sulforaphane modulates migration and EMT markers in breast cancer. Iran J Public Health 49:77

Bahadoran Z, Tohidi M, Nazeri P, Mehran M, Azizi F, Mirmiran P (2012) Broccoli sprouts improve insulin resistance in type 2 diabetes. Int J Food Sci Nutr 63:767–771

Bai Y, Cui W, Xin X, Miao M, Barati C, Zhang Q et al (2013) Sulforaphane prevents diabetic cardiomyopathy via Nrf2. J Mol Cell Cardiol 57:82–95

Baralić K, Živančević K, Marić Đ, Bozic D, Djordjevic AB, Miljaković EA et al (2023) Sulforaphane as strategy against toxic chemicals: colorectal carcinoma case study. Environ Res 227:115818

Basha SC, Babu KR, Madhu M, Gopinath C (2017) In vitro antidiabetic activity of sulforaphane. Pharmacol Toxicol Biomed Rep 3:2
Beaver LM, Kuintzle R, Buchanan A, Wiley MW, Glasser ST, Wong CP et al (2017) Long noncoding RNAs and sulforaphane in prostate cancer suppression. J Nutr Biochem 42:72–83
Bendary MM, Ali MA, Abdel Halim AS, Boufahja F, Chaudhary AA, Elkelish A et al (2024) Anti-virulence and anti-quorum sensing properties of sulforaphane. Front Pharmacol 15:1406653
Bernuzzi F, Maertens A, Saha S, Troncoso-Rey P, Ludwig T, Hiller K et al (2023) Sulforaphane rewires central metabolism to support antioxidant response. Redox Biol 67:102878
Bessetti RN, Litwa KA (2025) Broccoli for the brain: neuroprotective mechanisms of sulforaphane. Front Cell Neurosci 19:1601366
Borkowska J, Herman-Antosiewicz A, Hać A (2024) Isothiocyanates induce autophagy and inhibit protein synthesis via AMPK–mTORC1–S6K1. Eur J Nutr 64:46
Burnett JP, Lim G, Li Y, Shah RB, Lim R, Paholak HJ, Zhang T (2017) Sulforaphane enhances the anticancer activity of taxanes against triple negative breast cancer by killing cancer stem cells. Cancer Lett 394:52–64
Çakır I, Pan LP, Hadley CK, El-Gamal A, Fadel A, Elsayegh D, Ghamari-Langroudi M (2022) Sulforaphane reduces obesity by reversing leptin resistance. eLife 11:e67368
Calabrese EJ, Kozumbo WJ (2021) The phytoprotective agent sulforaphane prevents inflammatory degenerative diseases and age-related pathologies via Nrf2-mediated hormesis. Pharmacol Res 163:105283
Calcabrini C, Maffei F, Turrini E, Fimognari C (2020) Sulforaphane potentiates anticancer effects of doxorubicin and cisplatin and mitigates their toxic effects. Front Pharmacol 11:567
Casano AM, Peri F (2015) Microglia: multitasking specialists of the brain. Dev Cell 32(4):469–477
Cascajosa-Lira A, Prieto AI, Pichardo S, Jos A, Cameán AM (2024) Protective effects of sulforaphane against toxic substances and contaminants: a systematic review. Phytomedicine 130:155731
Cazalla E, Cuadrado A, García-Yagüe ÁJ (2024) Role of the transcription factor NRF2 in maintaining the integrity of the blood-brain barrier. Fluids Barriers CNS 21:1–29
Chegeni TN, Sarvi S, Moosazadeh M, Sharif M, Aghayan SA, Amouei A, Daryani A (2019) Is *toxoplasma gondii* a potential risk factor for Alzheimer's disease? A systematic review and meta-analysis. Microb Pathog 137:103751
Chen H, Lu J, Zou T, Teng Z, Qin Y, Wu R, Huang J (2025) Effects of sulforaphane on negative symptoms and cognitive impairments in chronic schizophrenia patients: a randomized double-blind trial. J Psychiatr Res 184:464–472
Cheng YM, Tsai CC, Hsu YC (2016) Sulforaphane, a dietary isothiocyanate, induces G2/M arrest in cervical cancer cells through cyclinB1 downregulation and GADD45β/CDC2 association. Int J Mol Sci 17(9):1530
Choi YJ, Lee WS, Lee EG, Sung MS, Yoo WH (2014) Sulforaphane inhibits IL-1β-induced proliferation of rheumatoid arthritis synovial fibroblasts and the production of MMPs, COX-2, and PGE2. Inflammation 37:1496–1503
Clark IA, Alleva LM, Vissel B (2010) The roles of TNF in brain dysfunction and disease. Pharmacol Ther 128(3):519–548
Clarke JD, Dashwood RH, Ho E (2008) Multi-targeted prevention of cancer by sulforaphane. Cancer Lett 269(2):291–304
Clulow JA, Storck EM, Lanyon-Hogg T, Kalesh KA, Jones LH, Tate EW (2017) Competition-based, quantitative chemical proteomics in breast cancer cells identifies new target profiles for sulforaphane. Chem Commun 53(37):5182–5185
Costa-Pérez A, Sánchez-Bravo P, Medina S, Domínguez-Perles R, García-Viguera C (2024) Bioaccessible organosulfur compounds in broccoli stalks modulate the inflammatory mediators involved in inflammatory bowel disease. Int J Mol Sci 25(2):800
Cox AG, Gurusinghe S, Abd Rahman R, Leaw B, Chan ST, Mockler JC, Wallace EM (2019) Sulforaphane improves endothelial function and reduces placental oxidative stress in vitro. Pregnancy Hypertens 16:1–10

Davidson R, Gardner S, Jupp O, Bullough A, Butters S, Watts L, Mithen R (2017) Isothiocyanates are detected in human synovial fluid following broccoli consumption and can affect the tissues of the knee joint. Sci Rep 7:3398

de Figueiredo S, Binda N, Nogueira-Machado J, Vieira-Filho S, Caligiorne R (2015) The antioxidant properties of organosulfur compounds (sulforaphane). Recent Pat Endocr Metab Immune Drug Discov 9:24–39

Deramaudt TB, Ali M, Vinit S, Bonay M (2020) Sulforaphane reduces intracellular survival of *Staphylococcus aureus* in macrophages through inhibition of JNK and p38 MAPK-induced inflammation. Int J Mol Med 45(6):1927–1941

Devi J, Thangam E (2012) Studies on antioxidant and antimicrobial activities of purified sulforaphane from *Brassica oleraceae* var rubra. J Pharm Res 5(7):3582–3584

Dong QQ, Wang QT, Wang L, Jiang YX, Liu ML, Hu HJ, Luo XG (2018) SMYD3-associated pathway is involved in the anti-tumor effects of sulforaphane on gastric carcinoma cells. Food Sci Biotechnol 27(4):1165–1173

Đurić A, Srdić-Rajić T, Božić D, Baralić K, Pavić A, Đukić-Ćosić D (2023) Sulforaphane affects morphological changes and cell viability of colon cancer cells: p53 mutation-dependent effect. In: 13th international congress of the Serbian Society of Toxicology and 1st TOXSEE regional conference. Udruženje toksikologa Srbije, Beograd, pp 180–180

Elbarbry F, Vermehren-Schmaedick A, Balkowiec A (2014) Modulation of arachidonic acid metabolism in the rat kidney by sulforaphane: implications for regulation of blood pressure. ISRN Pharmacol 2014:683508

Eren E, Tufekci KU, Isci KB, Tastan B, Genc K, Genc S (2018) Sulforaphane inhibits lipopolysaccharide-induced inflammation, cytotoxicity, oxidative stress, and miR-155 expression and switches to Mox phenotype through activating extracellular signal-regulated kinase 1/2–nuclear factor erythroid 2-related factor 2/antioxidant response element pathway in murine microglial cells. Front Immunol 9:36

Fahey JW, Stephenson KK, Wade KL, Talalay P (2013) Urease from *Helicobacter pylori* is inactivated by sulforaphane and other isothiocyanates. Biochem Biophys Res Commun 435(1):1–7

Fahey JW, Liu H, Batt H, Panjwani AA, Tsuji P (2025) Sulforaphane and brain health: from pathways of action to effects on specific disorders. Nutrients 17(8):1353

Feng Z, Wang T, Sun Y, Chen S, Hao H, Du W, Pang Y (2023) Sulforaphane suppresses paraquat-induced oxidative damage in bovine in vitro-matured oocytes through Nrf2 transduction pathway. Ecotoxicol Environ Saf 254:114747

Fernandez-Prades L, Brasal-Prieto M, Alba G, Martin V, Montserrat-de la Paz S, Cejudo-Guillen M, Lopez-Enriquez S (2023) Sulforaphane reduces the chronic inflammatory immune response of human dendritic cells. Nutrients 15(15):3405

Fix C, Carver-Molina A, Chakrabarti M, Azhar M, Carver W (2019) Effects of the isothiocyanate sulforaphane on TGF-β1-induced rat cardiac fibroblast activation and extracellular matrix interactions. J Cell Physiol 234(8):13931–13941

Gao L, Cheng D, Yang J, Wu R, Li W, Kong AN (2018) Sulforaphane epigenetically demethylates the CpG sites of the miR-9-3 promoter and reactivates miR-9-3 expression in human lung cancer A549 cells. J Nutr Biochem 56:109–115

Gao L, Du F, Wang J, Zhao Y, Liu J, Cai D, Zhang S (2021) Examination of the differences between sulforaphane and sulforaphene in colon cancer: a study based on next generation sequencing. Oncol Lett 22(4):1–10

Gasper AV, Al-Janobi A, Smith JA, Bacon JR, Fortun P, Atherton C, Mithen RF (2005) Glutathione S-transferase M1 polymorphism and metabolism of sulforaphane from standard and high-glucosinolate broccoli. Am J Clin Nutr 82:1283–1291

Ge Y, Ge Z, Tian F, Tai X, Chen D, Sun S, Ge X (2025) Sulforaphane potentiates the efficacy of chemoradiotherapy in glioblastoma by selectively targeting thioredoxin reductase 1. Cancer Lett 611:217429

Gorman A, Üçışık MH (2022) Lipopolysaccharide structure and the phenomenon of low endotoxin

Greaney AJ, Maier NK, Leppla SH, Moayeri M (2016) Sulforaphane inhibits multiple inflammasomes through an Nrf2-independent mechanism. J Leukoc Biol 99(1):189–199

Guerrero-Beltrán CE, Calderón-Oliver M, Pedraza-Chaverri J, Chirino YI (2012) Protective effect of sulforaphane against oxidative stress: recent advances. Exp Toxicol Pathol 64(5):503–508
Guo K, Wang L, Mahe J, Li L, Jiao S, Wang H, Hu X (2022) Effect of aqueous extract of seed of broccoli on inflammatory cytokines and *Helicobacter pylori* infection: a randomized, double-blind, controlled trial in patients without atrophic gastritis. Inflammopharmacology 30:1659–1668
Hać A, Brokowska J, Rintz E, Bartkowski M, Wegrzyn G, Herman-Antosiewicz A (2020) Mechanism of selective anticancer activity of isothiocyanates relies on differences in DNA damage repair between cancer and healthy cells. Eur J Nutr 59(4):1421–1432. https://doi.org/10.1007/s00394-019-01995-6
Haristoy X, Angioi-Duprez K, Duprez A, Lozniewski A (2003) Efficacy of sulforaphane in eradicating *helicobacter pylori* in human gastric xenografts implanted in nude mice. Antimicrob Agents Chemother 47(12):3982–3984
He C, Buongiorno LP, Wang W, Tang JC, Miceli N, Taviano MF, Bao Y (2021) The inhibitory effect of sulforaphane on bladder cancer cell depends on GSH depletion-induced by Nrf2 translocation. Molecules 26(16):4919
He L, Jiang H, Li Y, Zhang X, Sun W, Liu C, Wang C (2024) Sulforaphane-enriched extracts from broccoli exhibit antimicrobial activity against plant pathogens, promising a natural antimicrobial agent for crop protection. Biomolecules 14(3):352
Hei G, Smith RC, Li R, Ou J, Song X, Zheng Y, Wu R (2022) Sulforaphane effects on cognition and symptoms in first and early episode schizophrenia: a randomized double-blind trial. Schizophr Bull Open 3(1):sgac024
Hou TT, Yang HY, Wang W, Wu QQ, Tian YR, Jia JP (2018) Sulforaphane inhibits the generation of amyloid-β oligomer and promotes spatial learning and memory in Alzheimer's disease (PS1V97L) transgenic mice. J Alzheimer's Dis 62:1803–1813
Hu B, Cao P, Wang JH, Feng W, Zhang Y, Yang H (2024) Sulforaphane triggers Sirtuin 3-mediated ferroptosis in colorectal cancer cells via activating the AMP-activated protein kinase/mTOR signaling pathway. Hum Exp Toxicol 43:09603271241266106
Huang C, Wu J, Chen D, Jin J, Wu Y, Chen Z (2019) Effects of sulforaphane in the central nervous system. Eur J Pharmacol 853:153–168
Iida Y, Okamoto-Katsuyama M, Maruoka S, Mizumura K, Shimizu T, Shikano S, Gon Y (2021) Effective ferroptotic small-cell lung cancer cell death from SLC7A11 inhibition by sulforaphane. Oncol Lett 21(1):71
Jayakumar T, Chen WF, Lu WJ, Chou DS, Hsiao G, Hsu CY, Hsieh CY (2013) A novel antithrombotic effect of sulforaphane via activation of platelet adenylate cyclase: ex vivo and in vivo studies. J Nutr Biochem 24:1086–1095
Jiang X, Bai Y, Zhang Z, Xin Y, Cai L (2014) Protection by sulforaphane from type 1 diabetes-induced testicular apoptosis is associated with the up-regulation of Nrf2 expression and function. Toxicol Appl Pharmacol 279:198–210
Jiang Y, Kuang M, Deng C (2022) Effects of sulforaphane on apoptosis and PI3K/AKT/mTOR phosphorylation pathway in colorectal cancer cells. Acta Medica Mediterranea 38(1):227–231
Joshi H, Rani I, Sharma V, Sharma U, Dimri T, Kumar M, Tuli HS (2025) Sulforaphane: a natural organosulfur having potential to modulate apoptosis and survival signalling in cancer. Discov Oncol 16(1):1–14
Kaboli PJ, Khoshkbejari MA, Mohammadi M, Abiri A, Mokhtarian R, Vazifemand R, Xiao Z (2020) Targets and mechanisms of sulforaphane derivatives obtained from cruciferous plants with special focus on breast cancer—contradictory effects and future perspectives. Biomed Pharmacother 121:109635
Kan S, Wang J, Sun G (2018) Sulforaphane regulates apoptosis- and proliferation-related signaling pathways and synergizes with cisplatin to suppress human ovarian cancer. Int J Mol Med 42(5):2447–2458. https://doi.org/10.3892/ijmm.2018.3860
Kim DH, Sung B, Kang YJ, Hwang SY, Kim MJ, Yoon JH, Kim ND (2015) Sulforaphane inhibits hypoxia-induced HIF-1α and VEGF expression and migration of human colon cancer cells. Int J Oncol 47(6):2226–2232

Kim SH, Park HJ, Moon DO (2017) Sulforaphane sensitizes human breast cancer cells to paclitaxel-induced apoptosis by downregulating the NF-κB signaling pathway. Oncol Lett 13(6):4427–4432

Kiser C, Gonul CP, Olcum M, Genc S (2021) Inhibitory effects of sulforaphane on NLRP3 inflammasome activation. Mol Immunol 140:175–185. https://doi.org/10.1016/j.molimm.2021.10.014. Epub 2021 Oct 28. PMID: 34717147

Kong L, Wang H, Li C, Cheng H, Cui Y, Liu L, Zhao Y (2021) Sulforaphane ameliorates diabetes-induced renal fibrosis through epigenetic up-regulation of BMP-7. Diabetes Metab J 45(6):909–920

Konwiński RR, Haddad R, Chun JA, Klenow S, Larson SC, Haab BB, Furge LL (2004) Oltipraz, 3H-1,2-dithiole-3-thione, and sulforaphane induce overlapping and protective antioxidant responses in murine microglial cells. Toxicol Lett 153(3):343–355

Koolivand M, Ansari M, Piroozian F, Moein S, MalekZadeh K (2018) Alleviating the progression of acute myeloid leukemia (AML) by sulforaphane through controlling miR-155 levels. Mol Biol Rep 45:2491–2499

Krause K, Pyrczak-Felczykowska A, Karczewska M, Narajczyk M, Herman-Antosiewicz A, Szalewska-Pałasz A, Nowicki D (2021) Dietary isothiocyanates, sulforaphane and 2-phenethyl isothiocyanate, effectively impair *Vibrio cholerae* virulence. Int J Mol Sci 22(19):10187

Leiva-Castro C, Múnera-Rodríguez AM, Martínez-Bailén M, Carmona AT, López-Enríquez S, Palomares F (2025) Monovalent glycoconjugates of sulforaphane prevent inflammation induced by lipopolysaccharide in human dendritic cells by inhibiting NF κB signalling pathway. Br J Pharmacol 182:5157

Leng F, Hinz R, Gentleman S, Hampshire A, Dani M, Brooks DJ, Edison P (2023) Neuroinflammation is independently associated with brain network dysfunction in Alzheimer's disease. Mol Psychiatry 28(3):1303–1311

Leone A, Diorio G, Sexton W, Schell M, Alexandrow M, Fahey JW, Kumar NB (2017) Sulforaphane for the chemoprevention of bladder cancer: molecular mechanism targeted approach. Oncotarget 8(21):35412

Li X, Zhao Z, Li M, Liu M, Bahena A, Zhang Y, Liu G (2018) Sulforaphane promotes apoptosis, and inhibits proliferation and self-renewal of nasopharyngeal cancer cells by targeting STAT signal through miRNA-124-3p. Biomed Pharmacother 103:473–481

Li S, Yang H, Chen X (2019) Protective effects of sulforaphane on diabetic retinopathy: activation of the Nrf2 pathway and inhibition of NLRP3 inflammasome formation. Exp Anim 68(2):221–231

Licznerska B, Szaefer H, Krajka-Kuźniak V (2021) R-sulforaphane modulates the expression profile of AhR, ERα, Nrf2, NQO1, and GSTP in human breast cell lines. Mol Cell Biochem 476:525–533. https://doi.org/10.1007/s11010-020-03913-5

LiFang F, Yi T, LingLi Z (2019) Study on the inhibitory effects of sulforaphane on human gastric cancer cells and its mechanism. J Guangdong Pharm Univ 35(1):88–91

Lim S, Ahn JC, Lee EJ, Kim J (2020) Antiproliferation effect of sulforaphene isolated from radish (*Raphanus sativus* L.) seeds on A549 cells. Appl Biol Chem 63(1):75

Liu CM, Peng CY, Liao YW, Lu MY, Tsai ML, Yeh JC, Yu CC (2017) Sulforaphane targets cancer stemness and tumor initiating properties in oral squamous cell carcinomas via miR-200c induction. J Formos Med Assoc 116(1):41–48

Liu P, Wang W, Zhou Z, Smith AJO, Bowater RP, Wormstone IM, Bao Y (2018) Chemopreventive activities of sulforaphane and its metabolites in human hepatoma HepG2 cells. Nutrients 10(5):585. https://doi.org/10.3390/nu10050585

Liu P, Wang W, Tang J, Bowater RP, Bao Y (2019) Antioxidant effects of sulforaphane in human HepG2 cells and immortalised hepatocytes. Food Chem Toxicol 128:129–136

Livingstone TL (2022) Accumulation of Sulphur-containing dietary bioactives and the impact on the transcriptional signature of the prostate. Doctoral dissertation. University of East Anglia. https://ueaeprints.uea.ac.uk/id/eprint/84855/

Ma K, Miao L, Li B, Yu W, Liu F, Liu K, Yang Z (2024) Mechanism of action of Nrf2 and its related natural regulators in rheumatoid arthritis. J Orthop Surg Res 19(1):759

Mangla B, Javed S, Sultan MH, Kumar P, Kohli K, Najmi A, Ahsan W (2021) Sulforaphane: a review of its therapeutic potentials, advances in its nanodelivery, recent patents, and clinical trials. Phytother Res 35(10):5440–5458

Marshall SA, Young RB, Lewis JM, Rutten EL, Gould J, Barlow CK, Gulliver EL (2023) The broccoli-derived antioxidant sulforaphane changes the growth of gastrointestinal microbiota, allowing for the production of anti-inflammatory metabolites. J Funct Foods 107:105645

Martins T, Colaço B, Venâncio C, Pires MJ, Oliveira PA, Rosa E, Antunes LM (2018) Potential effects of sulforaphane to fight obesity. J Sci Food Agric 98(8):2837–2844

Masoumvand M, Ramezani E, Eshaghi Milasi Y, Baradaran Rahimi V, Askari VR (2025) New horizons for promising influences of sulforaphane in the management of metabolic syndrome: a mechanistic review. Naunyn Schmiedeberg's Arch Pharmacol 398(5):4933–4946

Mirmiran P, Bahadoran Z, Ghasemi A, Jeddi S, Azizi F (2017) High-sulforaphane broccoli sprout powder reduces serum nitric oxide metabolites in *Helicobacter pylori* infected patients. J Funct Foods 34:356–358

Mohamadi N, Baradaran Rahimi V, Fadaei MR, Sharifi F, Askari VR (2023) A mechanistic overview of sulforaphane and its derivatives application in diabetes and its complications. Inflammopharmacology 31(6):2885–2899

Mohanty S, Pal A, Konkimalla VB, Sudam Chandra SI (2018) Anti-inflammatory and anti-granuloma activity of sulforaphane, a naturally occurring isothiocyanate from broccoli (*Brassica oleracea*). Asian J Pharm Clin Res 11(6):411–416

Mokhtari RB, Qorri B, Baluch N, Sparaneo A, Fabrizio FP, Muscarella LA, Yeger H (2022). Correction:) next-generation multimodality of nutrigenomic cancer therapy: sulforaphane in combination with acetazolamide actively target bronchial carcinoid cancer in disabling the PI3K/Akt/mTOR survival pathway and inducing apoptosis. Oncotarget 13:968

Moon SJ, Jhun J, Ryu J, Kwon JY, Kim SY, Jung K, Min JK (2021) The anti-arthritis effect of sulforaphane, an activator of Nrf2, is associated with inhibition of both B cell differentiation and the production of inflammatory cytokines. PLoS One 16(2):e0245986

Mthembu SX, Mazibuko-Mbeje SE, Silvestri S, Orlando P, Nkambule BB, Muller CJ, Dludla PV (2025) Supplementation with aspalathin and sulforaphane protects cultured cardiac cells against dyslipidemia-associated oxidative damage. Metabolism Open 25:100346

Nicolas-Espinosa J, Yepes-Molina L, Carvajal M (2022) Bioactive peptides from broccoli stems strongly enhance regenerative keratinocytes by stimulating controlled proliferation. Pharm Biol 60(1):235–246

Nowicki D, Krause K, Karczewska M, Szalewska-Pałasz A (2021a) Evaluation of the anti-shigellosis activity of dietary isothiocyanates in *Galleria mellonella* larvae. Nutrients 13(11):3967

Nowicki D, Krause K, Szamborska P, Żukowska A, Cech GM, Szalewska-Pałasz A (2021b) Induction of the stringent response underlies the antimicrobial action of aliphatic isothiocyanates. Front Microbiol 11:591802

Ogunlade B, Adelakun S, Iteire K (2020) Sulforaphane response on aluminum-induced oxidative stress, alterations in sperm characterization and testicular histomorphometry in Wistar rats. Int J Reprod Biomed 18(8):611

Pacheco-Cano RD, Salcedo-Hernández R, Casados-Vázquez LE, Wrobel K, Bideshi DK, Barboza-Corona JE (2020) Class I defensins (BraDef) from broccoli (*Brassica oleracea* var. italica) seeds and their antimicrobial activity. World J Microbiol Biotechnol 36:1–12

Pan S, Lv Z, Wang R, Shu H, Yuan S, Yu Y, Shang Y (2022) Sepsis-induced brain dysfunction: pathogenesis, diagnosis, and treatment. Oxidative Med Cell Longev 1:1328729

Parfenova H, Liu J, Hoover DT, Fedinec AL (2020) Vasodilator effects of sulforaphane in cerebral circulation: a critical role of endogenously produced hydrogen sulfide and arteriolar smooth muscle KATP and BK channels in the brain. J Cereb Blood Flow Metab 40(10):1987–1996

Park J, Min JS, Kim B, Chae UB, Yun JW, Choi MS, Lee DS (2015) Mitochondrial ROS govern the LPS-induced pro-inflammatory response in microglia cells by regulating MAPK and NF-κB pathways. Neurosci Lett 584:191–196

Pastorek M, Simko V, Takacova M, Barathova M, Bartosova M, Hunakova L, Sedlak J (2015) Sulforaphane reduces molecular response to hypoxia in ovarian tumor cells independently of their resistance to chemotherapy. Int J Oncol 47(1):51–60

Peng X, Zhou Y, Tian H, Yang G, Li C, Geng Y, Wu W (2015) Sulforaphane inhibits invasion by phosphorylating ERK1/2 to regulate E-cadherin and CD44v6 in human prostate cancer DU145 cells. Oncol Rep 34(3):1565–1572

Pledgie-Tracy A, Sobolewski MD, Davidson NE (2007) Sulforaphane induces cell type-specific apoptosis in human breast cancer cell lines. Mol Cancer Ther 6(3):1013–1021

Pötschke V, Bereswill S, Heimesaat MM (2025) Antibacterial effects of sulforaphane—a phytonutrient derived from broccoli as promising candidate in the combat of bacterial infections. Eur J Microbiol Immunol 15:139

Qi T, Xu F, Yan X, Li S, Li H (2016) Sulforaphane exerts anti-inflammatory effects against lipopolysaccharide-induced acute lung injury in mice through the Nrf2/ARE pathway. Int J Mol Med 37(1):182–188

Rodova M, Fu J, Watkins DN, Srivastava RK, Shankar S (2012) Sonic hedgehog signaling inhibition provides opportunities for targeted therapy by sulforaphane in regulating pancreatic cancer stem cell self-renewal. PLoS One 7(9):e46083. https://doi.org/10.1371/journal.pone.0046083

Román J, Lagos A, Mahn A, Quintero J (2024) The effect of broccoli glucosinolates hydrolysis products on *Botrytis cinerea*: a potential new antifungal agent. Int J Mol Sci 25(14):7945

Royce SG, Licciardi PV, Beh RC, Bourke JE, Donovan C, Hung A, Karagiannis TC (2022) Sulforaphane prevents and reverses allergic airways disease in mice via anti-inflammatory, antioxidant, and epigenetic mechanisms. Cell Mol Life Sci 79(11):579

Ruhee RT, Suzuki K (2020) The integrative role of sulforaphane in preventing inflammation, oxidative stress and fatigue: a review of a potential protective phytochemical. Antioxidants 9(6):521

Ruhee RT, Suzuki K (2024) The immunomodulatory effects of sulforaphane in exercise-induced inflammation and oxidative stress: a prospective nutraceutical. Int J Mol Sci 25(3):1790

Russo M, Spagnuolo C, Russo GL, Skalicka-Woźniak K, Daglia M, Sobarzo-Sánchez E, Nabavi SM (2018) Nrf2 targeting by sulforaphane: a potential therapy for cancer treatment. Crit Rev Food Sci Nutr 58(8):1391–1405

Rutz J, Thaler S, Maxeiner S, Chun FK, Blaheta RA (2020) Sulforaphane reduces prostate cancer cell growth and proliferation in vitro by modulating the CDK-cyclin axis and expression of the CD44 variants 4, 5, and 7. Int J Mol Sci 21(22):8724. https://doi.org/10.3390/ijms21228724

Saito A, Ishikawa S, Yang K, Sawa A, Ishizuka K (2025) Sulforaphane as a potential therapeutic agent: a comprehensive analysis of clinical trials and mechanistic insights. J Nutr Sci 14:e65

Santín-Márquez R, Alarcón-Aguilar A, López-Diazguerrero NE, Chondrogianni N, Königsberg M (2019) Sulforaphane—role in aging and neurodegeneration. Geroscience 41:655–670

Schepici G, Bramanti P, Mazzon E (2020) Efficacy of sulforaphane in neurodegenerative diseases. Int J Mol Sci 21(22):8637

Shahlaei M, Saeidifar M, Zamanian A (2022) Sustained release of sulforaphane by bioactive extracellular vesicles for neuroprotective effect on chick model. J Biomed Mater Res B Appl Biomater 110(12):2636–2648

Shawky NM, Segar L (2018) Sulforaphane improves leptin responsiveness in high-fat high-sucrose diet-fed obese mice. Eur J Pharmacol 835:108–114

Silva BL, Simão G, Campos CD, Monteiro CR, Bueno LR, Ortis GB, Fernandes ES (2022) In silico and in vitro analysis of sulforaphane anti-*Candida* activity. Antibiotics 11(12):1842

Silva-Palacios A, Ostolga-Chavarría M, Sánchez-Garibay C, Rojas-Morales P, Galván-Arzate S, Buelna-Chontal M, Zazueta C (2019) Sulforaphane protects from myocardial ischemia-reperfusion damage through the balanced activation of Nrf2/AhR. Free Radic Biol Med 143:331–340

Singh A, Ansari VA, Mahmood T, Ahsan F, Wasim R (2022) Neurodegeneration: microglia: NF-κB signaling pathways. Drug Res 72(09):496–499

Socała K, Nieoczym D, Kowalczuk-Vasilev E, Wyska E, Wlaź P (2017) Increased seizure susceptibility and other toxicity symptoms following acute sulforaphane treatment in mice. Toxicol Appl Pharmacol 326:43–53

Song K, Li Y, Zhang H, An N, Wei Y, Wang L, Gao Y (2020) Oxidative stress-mediated blood-brain barrier (BBB) disruption in neurological diseases. Oxidative Med Cell Longev 1:4356386

Su X, Jiang X, Meng L, Dong X, Shen Y, Xin Y (2018) Anticancer activity of sulforaphane: the epigenetic mechanisms and the Nrf2 signaling pathway. Oxidative Med Cell Longev 2018:5438179

Subedi L, Lee JH, Yumnam S, Ji E, Kim SY (2019) Anti-inflammatory effect of sulforaphane on LPS-activated microglia potentially through JNK/AP-1/NF-κB inhibition and Nrf2/HO-1 activation. Cells 8(2):194

Sun Y, Zhou S, Guo H, Zhang J, Ma T, Zheng Y, Cai L (2020) Protective effects of sulforaphane on type 2 diabetes-induced cardiomyopathy via AMPK-mediated activation of lipid metabolic pathways and NRF2 function. Metabolism 102:154002

Treasure K, Harris J, Williamson G (2023) Exploring the anti-inflammatory activity of sulforaphane. Immunol Cell Biol 101(9):805–828

Treasure K, Harris J, Williamson G (2024) Exploring the anti-inflammatory activity of the plant-derived phytochemical sulforaphane from cruciferous vegetables. Proc Nutr Soc 83(OCE1):E177

Türker NP, Bağcı U, Onsekizoğlu-Bağcı P (2019) Investigation of the anticancer and proliferative effect of broccoli extract on Du145 prostate cancer and MEF healthy fibroblast cell lines. Int J Innov Approaches Agric Res 3(4):550–556

Uddin MS, Al Mamun A, Jakaria M, Thangapandiyan S, Ahmad J, Rahman MA, Aleya L (2020) Emerging promise of sulforaphane-mediated Nrf2 signaling cascade against neurological disorders. Sci Total Environ 707:135624

Von Bernhardi R, Eugenín-von Bernhardi L, Eugenín J (2015) Microglial cell dysregulation in brain aging and neurodegeneration. Front Aging Neurosci 7:124

Wang X, Li Y, Dai Y, Liu Q, Ning S, Liu J, Li Z (2016) Sulforaphane improves chemotherapy efficacy by targeting cancer stem cell-like properties via the miR-124/IL-6R/STAT3 axis. Sci Rep 6(1):36796

Wang DX, Zou YJ, Zhuang XB, Chen SX, Lin Y, Li WL, Lin ZQ (2017) Sulforaphane suppresses EMT and metastasis in human lung cancer through miR-616-5p-mediated GSK3β/β-catenin signaling pathways. Acta Pharmacol Sin 38(2):241–251

Wang F, Sun Y, Huang X, Qiao C, Zhang W, Liu P, Wang M (2021) Sulforaphane inhibits self-renewal of lung cancer stem cells through the modulation of sonic hedgehog signaling pathway and polyhomeotic homolog 3. AMB Express 11:1–10

Wang X, Chen X, Zhou W, Men H, Bao T, Sun Y, Cai L (2022) Ferroptosis is essential for diabetic cardiomyopathy and is prevented by sulforaphane via AMPK/NRF2 pathways. Acta Pharm Sin B 12(2):708–722

Wang P, Wang Z, Jin X, Zhang M, Shen M, Li D (2025) Oral sulforaphane intervention protects against diabetic cardiomyopathy in db/db mice: focus on cardiac lipotoxicity and substrate metabolism. Antioxidants 14(5):603

Welin-Neilands J, Svensater G (2007) Acid tolerance of biofilm cells of *Streptococcus mutans*. Appl Environ Microbiol 73(17):5633–5638

Wu W, Peng G, Yang F, Zhang Y, Mu Z, Han X (2019) Sulforaphane has a therapeutic effect in an atopic dermatitis murine model and activates the Nrf2/HO-1 axis. Mol Med Rep 20(2):1761–1771

Xie Y, Zhang G, Li X, Huang D (2016) Sulforaphane suppressed proliferation of lung cancer stem cells in vitro. Cancer Res Prev Treat 43(7):555–559

Xie C, Zhu J, Jiang Y, Chen J, Wang X, Geng S, Meng Z (2019) Sulforaphane inhibits the acquisition of tobacco smoke-induced lung cancer stem cell-like properties via the IL-6/ΔNp63α/Notch axis. Theranostics 9(16):4827

Xin Y, Bai Y, Jiang X, Zhou S, Wang Y, Wintergerst KA, Cai L (2018) Sulforaphane prevents angiotensin II-induced cardiomyopathy by activation of Nrf2 via stimulating the Akt/GSK-3β/Fyn pathway. Redox Biol 15:405–417

Xu Y, Liu C, Chen S, Ye Y, Guo M, Ren Q, Chen L (2014) Activation of AMPK and inactivation of Akt result in suppression of mTOR-mediated S6K1 and 4E-BP1 pathways leading to neuronal cell death in in vitro models of Parkinson's disease. Cell Signal 26(8):1680–1689

Xu Y, Huang X, Huangfu B, Hu Y, Xu J, Gao R et al (2023) Sulforaphane ameliorates nonalcoholic fatty liver disease induced by high-fat and high-fructose diet via LPS/TLR4 in the gut–liver axis. Nutrients 15(3):743

Xue M, Qian Q, Adaikalakoteswari A, Rabbani N, Babaei-Jadidi R, Thornalley PJ (2008) Activation of NF-E2-related factor-2 reverses biochemical dysfunction of endothelial cells induced by hyperglycemia linked to vascular disease. Diabetes 57(10):2809–2817

Yagishita Y, Fahey JW, Dinkova-Kostova AT, Kensler TW (2019) Broccoli or sulforaphane: is it the source or dose that matters? Molecules 24(19):3593

Yan Y, Zhou Y, Li J, Zheng Z, Hu Y, Li L, Wu W (2021) Sulforaphane downregulated fatty acid synthase and inhibited microtubule-mediated mitophagy leading to apoptosis. Cell Death Dis 12(10):917

Yang D, Han B, Baiyun R, Lv Z, Wang X, Li S, Zhang Z (2020) Sulforaphane attenuates hexavalent chromium-induced cardiotoxicity via the activation of the Sesn2/AMPK/Nrf2 signaling pathway. Metallomics 12(12):2009–2020

Yu M, Chen Y, Dong S, Chen Z, Jiang X, Wang Y, Zhang L (2025) Sulforaphane as a promising anti-caries agent: inhibitory effects on *Streptococcus mutans* and caries control in a rat model. Front Microbiol 15:1427803

Yunusoğlu O, Koyuncu E, Kalfa İ (2025) Sulforaphane in the therapeutic targeting of CNS disorders: current evidence and perspectives. Experimed 15(3):201–216

Zarzycka M, Kotula-Balak M, Gil D (2026) Inhibiting autophagy enhances anti-cancer properties of sulforaphane. Sci Rep 16:5296

Zhang L, Tu Y (2018) Effects of sulforaphane on the apoptosis in human gastric cancer cells via regulating PI3K/Akt signaling pathway. Int J Tradit Chin Med:628–632

Zhang R, Zhang J, Fang L, Li X, Zhao Y, Shi W, An L (2014) Neuroprotective effects of sulforaphane on cholinergic neurons in mice with Alzheimer's disease-like lesions. Int J Mol Sci 15(8):14396–14410

Zhang J, Dong Y, Zhou M, Wo X, Niu S, Shao E et al (2022) Sulforaphane protects myocardium from ischemia-reperfusion injury by regulating CaMKIIN2 and CaMKIIδ. Biochem Biophys Res Commun 605:119–126

Zhang J, Yang C, Zhang S, Cao T, Peng R, Guo W, Huang T (2023) Sulforaphane reverses Aβ fiber-mediated M1 type microglia polarization and neuroinflammation-mediated necroptosis of neural stem cells by downregulating the MAPK/NF-κB signaling pathways. J South Med Univ 43(12):2132–2138

Zhao Y, Shang S, Song Y, Li T, Han M, Qin Y, Tang B (2022) Sulforaphane kills *Mycobacterium tuberculosis* H37Ra and *Mycobacterium smegmatis* mc^2155 through a reactive oxygen species dependent mechanism. J Microbiol 60(11):1095–1105

Zhao Z, Chen Q, Qiao X, Wang J, Ali ZTA, Li J, Yin L (2025a) Sulforaphane in cancer precision medicine: from biosynthetic origins to multiscale mechanisms and clinical translation. Front Immunol 16:1702860

Zhao L, Li J, Dang Y, Fisher D, Hien NTT, Musabaev E, Zhao L (2025b) Protective role of sulforaphane in lipid metabolism-related diseases. Mol Biol Rep 52(1):241

Zheng W, Li X, Zhang T, Wang J (2022) Biological mechanisms and clinical efficacy of sulforaphane for mental disorders. Gen Psychiatr 35(2):e100700

Zhou Q, Chen B, Wang X, Wu L, Yang Y, Cheng X, Cao P (2016) Sulforaphane protects against rotenone-induced neurotoxicity in vivo: involvement of the mTOR, Nrf2 and autophagy pathways. Sci Rep 6(1):32206

Zhu Y, Hu H, Huang K, Sun X, Li M (2022) Sulforaphane inhibits proliferation and apoptosis of colorectal cancer cells by down-regulating the COX-2/PKB/GSK-3β signaling pathway. Indian J Pharm Sci 84(1)

Żuryń A, Litwiniec A, Safiejko-Mroczka B, Klimaszewska-Wiśniewska A, Gagat M, Krajewski A, Grzanka D (2016) The effect of sulforaphane on the cell cycle, apoptosis and expression of cyclin D1 and p21 in the A549 non-small cell lung cancer cell line. Int J Oncol 48(6):2521–2533

Żuryń A, Krajewski A, Klimaszewska-Wiśniewska A, Grzanka A, Grzanka D (2019) Expression of cyclin B1, D1 and K in non-small cell lung cancer H1299 cells following treatment with sulforaphane. Oncol Rep 41(2):1313–1323

Chapter 4
Clinical Applications, Formulation Strategies, and Future Directions of Sulforaphane

Abstract Sulforaphane (SFN) has demonstrated broad therapeutic potential across inflammation-related conditions, including cancer, cardiovascular disease, and neurodegenerative disorders, yet its clinical translation remains constrained by chemical instability and highly variable oral bioavailability. Conversion of glucoraphanin to active SFN depends on myrosinase activity and individual gut microbiota composition, resulting in inconsistent absorption and efficacy and contributing to its current status as a dietary supplement rather than an approved drug. This chapter synthesises clinical trial data across diverse indications—neurological and psychiatric disorders, cardiovascular disease, chronic kidney disease, type 2 diabetes, and dermatological conditions—and compares them with the preclinical evidence base, revealing a pronounced translational gap: oncology dominates laboratory research, whereas clinical studies follow a distinct thematic distribution shaped largely by epidemiological hypotheses. Variability in study design, dosing, formulation, and endpoints further complicates cross-study comparison, leaving open the question of whether preclinical mechanisms translate equivalently to humans. Advanced formulation strategies developed over the past 5 years to overcome SFN's instability and rapid metabolism are also reviewed, including nanoemulsions, lipid and polymeric nanoparticles, cyclodextrin inclusion complexes, microencapsulation, and biological carriers such as broccoli-derived nanovesicles, alongside protein- and polysaccharide-based encapsulation. Future directions emphasise stereochemical characterisation of SFN enantiomers and exploration of synergistic interactions with conventional pharmacological agents, supporting continued development from bench to bedside.

Keywords Clinical translation · Bioavailability · Formulation · Drug delivery systems

K. Sikorska-Zimny, *Sulforaphane in Human Health*, SpringerBriefs in Food, Health, and Nutrition, https://doi.org/10.1007/978-3-032-34331-4_4

4.1 Clinical Trials in a Dish

Owing to its distinctive and therapeutically desirable properties, sulforaphane is currently being investigated for its potential applications in medical treatments. From a pharmacological perspective, SNF complies with multiple criteria for drug-like behaviour, including the following:

- Lipinski's Rule of Five (RO5): This rule evaluates a compound's absorption, distribution, metabolism, and excretion (ADME) to determine its potential bioactivity in the human body (Zhang and Wilkinson 2007).
- Ghose Filter: This criterion predicts the drug-like properties of a compound (Pinlac et al. 2021).
- Veber's Rule: This rule assesses oral bioavailability, indicating that compounds with ten or fewer rotatable bonds and a polar surface area no greater than 140 $Å^2$ exhibit favourable oral bioavailability (Veber et al. 2002).

In Clinical Trials in a Dish (CITD), a method allowing the evaluation of medical therapies for safety and efficacy using patient-derived cells in human clinical trials (Fermini et al. 2018), sulforaphane is being tested for its potential therapeutic effects in various disorders and diseases (PubChem n.d.; Drug Bank n.d.) (Table 4.1).

4.2 Translational Challenges and Limitations

The clinical trials listed in the table vary considerably in terms of target disease, study phase, study population, and time. Most studies are in the early or mid-stage (phase 1-2), emphasising the ongoing evaluation of safety, feasibility, and preliminary efficacy in various clinical settings. Because SFN is known for its anticancer effects, most studies have focused on oncology applications, particularly cancer chemoprevention and adjuvant treatment strategies (Fig. 4.1). Bladder, lung, and prostate cancers are the most frequently studied malignancies in this category of cancer. A randomised phase 2 clinical trial evaluating the efficacy of sulforaphane in the chemoprevention of bladder cancer represents a key step toward translating preclinical evidence of carcinogen detoxification and apoptosis induction in humans. Similarly, studies investigating lung cancer chemoprevention in former smokers aim to assess whether enhancing endogenous antioxidant mechanisms with sulforaphane reduces the long-term risk of carcinogenicity. Prostate cancer is the subject of numerous studies, including those on the accumulation of bioactive substances from the diet and interventional studies in patients with relapsed disease. These studies reflect the interest in the potential of sulforaphane to modulate androgen signalling, inflammation, and oxidative stress, which are crucial mechanisms for prostate cancer progression.

Beyond oncology, sulforaphane has been extensively studied in relation to organ dysfunction, particularly in the brain, heart, kidneys, and lungs of rats. Neurological

Table 4.1 Clinical trials for sulforaphane

	Target	Name	CITD no.	Phase	Date
Cancer	Bladder	Randomized, Phase II Clinical Trial of Sulforaphane in Bladder Cancer Chemoprevention	NCT03517995	2	28.05.2020
	Lungs	Clinical Trial of Lung Cancer Chemoprevention with Sulforaphane in Former Smokers	NCT03232138	2	17.12.2024
	Prostate	Accumulation of Dietary Bioactives and Prostate Cancer	NCT04046653	N/A	18.11.2020
	Prostate	Sulforaphane in Treating Patients with Recurrent Prostate Cancer	NCT01228084	2	24.08.2017
Internal organ dysfunctions	Brain damage	Effects of Sulforaphane on Cognitive Function in Patients with Frontal Brain Damage	NCT04252261	2	19.05.2020
	Cardiac	Protective Effects of the Nutritional Supplement Sulforaphane on Doxorubicin-Associated Cardiac Dysfunction	NCT03934905	1/2	20.11.2024
	Cardiac	Prevention of Age-associated Cardiac and Vascular Dysfunction Using Avmacol ES	NCT05408559	1/2	06.11.2024
	Kidneys	Safety, Feasibility and Efficacy of Sulforaphane (Avmacol Extra Strength) in Chronic Kidney Disease	NCT05797506	2	17.12.2024
	Kidneys	Safety, Feasibility and Tolerability of Sulforaphane in Children with Chronic Kidney Disease	NCT05653492	2	2024.09.19 19.02.2024
	Kidneys	Study of Sulphoraphane in Chronic Kidney Disease	NCT05153174	1	12.08.2024
	Nose	Broccoli Sprout Extract Effects on Allergic Inflammation in the Nose	NCT00882115	-	Comp.
	Pulmonary	Broccoli Sprout Extracts Trial to See if NRF2 Is Enhanced by Sulforaphane Treatment in Patients With COPD	NCT01335971	2	Comp.

(continued)

Table 4.1 (continued)

	Target	Name	CITD no.	Phase	Date
Disorders of human cognition, affect and behaviour (Crespi and Badcock 2008)	Alzheimer's disease	Effects of Sulforaphane in Patients with Prodromal to Mild Alzheimer's Disease	NCT04213391	N/A	12.05.2020
	Attention deficit disorder	Study of the Efficacy of Sulforaphane in Children Aged 6 to 12 With Attention Deficit Disorder with or Without Hyperactivity	NCT06594536	N/A	19.09.2024
	Autism	Sulforaphane in a New Jersey (NJ) Population of Individuals with Autism	NCT02677051	2	18.11.2024
	Autism	Sulforaphane Treatment of Children with Autism Spectrum Disorder (ASD)	NCT02561481	1/2	17.12.2020
	Autism	A 12-weeks Study to Evaluate Sulforaphane in Treatment of Autism Spectrum Disorder	NCT02879110	N/A	06.03.2019
	Autism	Sulforaphane for the Treatment of Young Men with Autism Spectrum Disorder	NCT02909959	2	04.06.2020
	Depression	A 12-weeks Study to Evaluate Sulforaphane in the Treatment of Depression	NCT04246905	2	20.08.2020
	Depressive disorder	A Comparative Study on Efficacy and Safety of add-on Sulforaphane or rTMS to Escitalopram for Major Depressive Disorder with Poor Response to Initial Treatment	NCT05145270	4	06.12.2021
	Parkinson diseases	A 6-month Study to Evaluate Sulforaphane Effects in PD Patients	NCT05084365	2	06.07.2023
	Psychosis	Decreasing Risk of Psychosis by Sulforaphane (DROPS Trial)	NCT03932136	3	31.10.2023
	Schizophrenia	A 6-month Study to Evaluate Sulforaphane Effects in Schizophrenia Patients	NCT04521868	2	26.10.2024
	Schizophrenia	A 6-month Study to Evaluate Sulforaphane add-on Effects in Treatment of Schizophrenia	NCT02880462	2	24.03.2024

Body diseases	Diabetes	Clinical Trial with Broccoli Sprout Extract to Patients with Type 2 Diabetes	NCT02801448	2	19.01.2022
	Osteoarthritis	Broccoli In Osteoarthritis	NCT03878368	N/A	07.05.2022
	Suppressed HIV	Sulforaphane Use to Effect Inflammatory and Metabolic Changes in Virally Suppressed HIV Patients	NCT05224492	4	07.02.2022
Skin	Nevi-precursor lesions	Pilot Study Evaluating Sulforaphane in Atypical Nevi-Precursor Lesions	NCT01568996	0	Comp.
	Skin aging	Effect of Topical Sulforaphane on Skin Aging and With Ultraviolet and Visible Light Exposure	NCT03730649	1	19.09.2024
	Skin aging	Effect of Topical Sulforaphane on Skin Fragility Seen in Skin Aging and With Ultraviolet Exposure	NCT03126539	2	29.06.2023

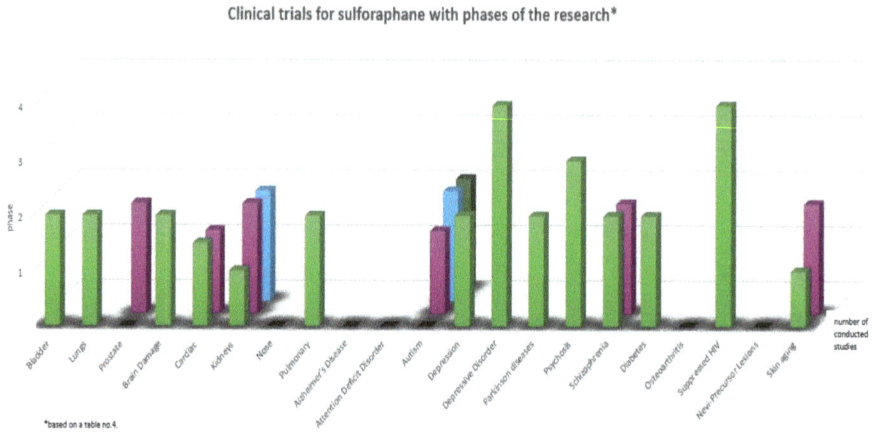

Fig. 4.1 Comparison of the number of studies using SFN in relation to specific diseases and their corresponding clinical phases

impairment is the subject of a Phase 2 study examining the effects of sulforaphane on cognitive function in patients with frontal lobe brain damage, suggesting a potential neuroprotective role related to redox homeostasis and inflammation. Cardiology studies include the evaluation of sulforaphane supplements for protection against doxorubicin-induced cardiotoxicity and the prevention of age-related cardiac and vascular dysfunction. Phase 1-2 studies emphasise safety and early efficacy, reflecting preclinical evidence that sulforaphane may help preserve mitochondrial function and mitigate oxidative damage in heart tissue. Kidney disease is a particularly active area of research. Numerous studies have evaluated the safety, tolerability, and efficacy of sulforaphane in adults and children with chronic kidney disease (CKD). Re-examination of phase 2 and 1 studies in this area underscores the validity of investigating sulforaphane as a nephroprotective agent (potentially by reducing inflammation, fibrosis, and oxidative stress). Lung and nasal inflammation have also been explored, including studies on broccoli sprout extracts in allergic rhinitis and chronic obstructive pulmonary disease (COPD). These studies often focus on enhancing Nrf2 activity as a biomarker-driven outcome, linking molecular mechanisms directly to clinical endpoints. The largest number of studies (12 of 30) in the table are devoted to cognitive, affective, and behavioural disorders in humans, reflecting the growing interest in the effects of sulforaphane on brain function. Alzheimer's disease, attention deficit disorder, autism spectrum disorder (ASD), depression, schizophrenia, psychosis, and Parkinson's disease were represented. In Alzheimer's disease, studies have evaluated the effects of sulforaphane in patients with prodromal and mild stages, aiming for early intervention in neurodegenerative processes. Studies on attention deficit disorder have focused on paediatric populations, highlighting the compound's potential role in neurodevelopmental modulation. Autism spectrum disorders are particularly well represented, including in children and young adults. These studies build on previous findings, suggesting improved behavioural and social functioning with sulforaphane administration. Similarly, studies on schizophrenia and psychosis are exploring sulforaphane as an adjunctive therapy, potentially targeting oxidative stress and neuroinflammation, pathophysiological processes increasingly recognised in severe mental illness. Mood disorders, including major depressive disorder, are being treated in both standalone and complementary interventional studies, some of which are in advanced phase 4.

The systemic effects of sulforaphane are being further explored in studies on type 2 diabetes, osteoarthritis, and inflammatory and metabolic changes in virally suppressed HIV patients. These studies reflect the broad impact of this compound on metabolism, including insulin sensitivity, inflammatory pathways, and immune regulation. Dermatological applications include pilot studies of atypical naevi as precursor lesions and studies on the topical use of sulforaphane in the treatment of skin ageing and photoageing. Notably, topical use of SFN leverages its antioxidant properties to protect against skin damage induced by ultraviolet radiation and visible light.

Among the more than 40 available studies analysed, 34 focused on anticancer therapy (Table 3.1). Notwithstanding the greatest interest in clinical trials focusing on its potential applications in disorders related to human cognition, affect, and behaviour, 12 out of 30 Clinical Trials in a Dish (CITD) studies investigated these aspects (Table 4.1). When comparing the data presented in both tables, a clear discrepancy was observed between the primary focus of laboratory research (cancer) and the intended or proposed therapeutic applications (cognitive impairment) in studies involving patients and volunteers.

The former primarily focuses on treating diseases associated with brain dysfunction, whereas laboratory studies primarily investigate the effects of these drugs on cancer. This poses a significant translational challenge; while preclinical research remains primarily focused on cancer biology, clinical trials increasingly address psychiatric, neurological, and metabolic disorders. This discrepancy illustrates a broader problem in translational medicine: strong evidence from laboratory models does not always translate into a coherent clinical research program. In some cases, clinical interest, particularly in psychiatric disorders, appears to be driven more by epidemiological observations and theoretical mechanisms than by extensive preclinical data. Although the reasons for this discrepancy cannot be definitively determined, these differences may stem from the knowledge of sulforaphane's specific mechanisms of action (e.g. anti-inflammatory properties) and the ability to observe these effects in cellular and animal models (e.g. inhibition of tumour growth). These factors pose significant challenges in human trials due to external variables such as cancer type, concomitant therapies, and the composition of the control and experimental groups.

Both tables present complementary but fundamentally different levels of evidence regarding the biological and therapeutic effects of sulforaphane, reflecting different stages of scientific research and translational development. Table 4.1 consists exclusively of registered clinical trials covering a broad spectrum of diseases and physiological systems, whereas Table 3.1 aggregates in vitro, ex vivo, and animal studies focusing on molecular targets, cellular responses, and biochemical pathways affected by sulforaphane. A parallel analysis of these datasets revealed both strong thematic convergence and significant methodological and conceptual divergence, illustrating how knowledge from preclinical studies has influenced clinical hypotheses while also revealing gaps between experimental results and clinical validation. The main point of contrast between the two tables is the level of biological organisation. The clinical trials table focuses on the results of whole-organism studies in human populations, assessing safety, feasibility, tolerability, and therapeutic or preventive efficacies. The endpoints in these studies were necessarily integrated, often including clinical symptoms, functional outcomes, disease progression, or validated biomarkers. In contrast, preclinical studies are dominated by reductionist experimental systems, including immortalised human cancer cell lines, primary cells, enzyme-based assays, rodent models, and microbial cultures. These studies have examined the effects of sulforaphane at the molecular and cellular levels, such as the modulation of transcription factors, signalling cascades, epigenetic regulators, oxidative stress markers, apoptosis-related proteins, and metabolic enzymes. Consequently, the two tables differ fundamentally in the nature of their results: the former emphasises the specificity of preclinical studies, whereas the latter emphasises clinical relevance. Despite this difference in scale, there is a significant overlap in the areas of interest, particularly in oncology. As shown in Table 4.1, cancer studies primarily focus on bladder, lung, and prostate cancers, often as chemopreventive or adjunctive therapies. These studies are typically conducted in Phases 1-3 and focus on safety, tolerability, and surrogate biomarkers rather than final survival outcomes. In contrast, Table 3.1 covers a much broader range of cancer types, including breast, colon, gastric, lung, prostate, liver, bladder, leukaemia, glioblastoma multiforme, oral squamous cell carcinoma, and bronchial carcinoid tumours. This discrepancy reflects the exploratory nature of preclinical studies, in which sulforaphane was evaluated in multiple tumour models to identify susceptible tumour types and molecular pathways.

The table of preclinical studies provides detailed insights into the cellular processes affected by sulforaphane. Many of these are conceptually aligned with clinical trial designs but are not directly measured in human studies. In vitro and animal experiments have consistently demonstrated that sulforaphane inhibits cancer cell proliferation, induces apoptosis, promotes cell cycle arrest, inhibits migration and invasion, and disrupts cancer stem cell self-renewal. These effects are mediated through the modulation of signalling pathways, such as NF-κB, PI3K/Akt/mTOR, β-catenin, ERK, SHh, and ferroptosis-related signalling, as well as through epigenetic mechanisms involving histone deacetylases, DNA methyltransferases, and microRNAs. However, clinical trials rarely directly assess these molecular endpoints and present results based on systemic biomarkers or clinical outcomes. This highlights a translational gap: while preclinical studies offer detailed explanations of the anticancer effects of sulforaphane, clinical trials often lack a clear answer to confirm whether these mechanisms operate in humans.

Another significant difference between the tables is the dose and formulation. In preclinical studies, sulforaphane is typically used at precisely defined concentrations, often ranging from submicromolar to high micromolar, dissolved in solvents such as DMSO, or administered as purified enantiomers. In vitro studies have provided fundamental molecular evidence for SFN's biological activity of SFN, particularly in oncology. A surprisingly wide range of SFN concentrations has been used in experimental studies, from as low as 0.625 μM to as high as 200 μM. However, a significant methodological limitation should be considered: many in vitro studies use SFN concentrations in the range of 40–200 μM, which far exceeds physiologically achievable levels. In humans, plasma concentrations following dietary intake rarely exceed 2–3 μM. Some studies have used broccoli extracts, juices, or racemic mixtures, and dosing remains strictly controlled and is administered directly to cells or animals. Another concern is the potential toxicity of SFN, which has been observed at high concentrations used in these studies.

Another obstacle to the translational development of SFN is the so-called dosage and formulation gap. In preclinical experiments, SFN is often dissolved in DMSO and administered at precisely controlled concentrations that may exceed physiological levels. In clinical trials, SFN is typically administered in formulations such as broccoli sprout extracts or standardised supplements, where dosing must consider bioavailability, metabolic conversion from glucoraphanin, and rapid elimination of the compound from the body. Despite compelling preclinical evidence, the clinical use of sulforaphane faces challenges related to its bioavailability and metabolic stability (SFN exhibits a relatively short half-life and rapid metabolism in humans, with its plasma concentration peaking approximately 1 hour after ingestion; strategies to enhance its stability and bioavailability, such as encapsulation, microencapsulation, and the addition of exogenous myrosinase, are currently being explored to optimise therapeutic effects). Although sulforaphane research has been ongoing for decades and is characterised by multi-level biological complexity, a degree of conceptual coherence in this field is maintained by consistently emphasising oxidative stress and inflammation as the central biological mechanisms. Continuous integration of findings from preclinical studies with well-designed clinical trials is essential to fully exploit the therapeutic potential of sulforaphane.

The concentrations used often exceed those achievable through diet alone. However, the clinical studies summarised in Table 4.1 utilised preparations intended for human use, such as broccoli sprout extracts, standardised dietary supplements, or commercial preparations (e.g. Avmacol). It is also important to consider that dosing in clinical trials must consider bioavailability, metabolism, compliance, and safety, leading to significantly lower and more variable systemic exposure. This discrepancy complicates the direct translation of dose-response relationships from preclinical studies to clinical settings and highlights the importance of pharmacokinetic considerations, which are largely not addressed in in vitro studies.

The tables also differ significantly in their treatment of non-oncological conditions, demonstrating the diversification of sulforaphane research over time. The clinical table covers a broad range of neurological, psychiatric, metabolic, cardiovascular, renal, and pulmonary diseases. These studies reflect the growing interest in sulforaphane as a systemic cytoprotective and neuroprotective agent, extending beyond its initial identification as a chemopreventive agent. In contrast, the preclinical table, while still strongly focused on cancer, encompasses a more limited but conceptually harmonised set of non-cancer models, such as insulin resistance, anti-diabetic enzyme inhibition, inflammatory responses, endothelial dysfunction, acute lung injury, bacterial infections, and cellular processes associated with neurodegeneration.

The studies presented in Table 3.1 often investigate fundamental mechanisms, such as redox balance, inflammatory signalling, mitochondrial function, and autophagy, which are relevant to many of the disease categories included in the clinical table. A particularly strong convergence point between both tables is the emphasis on oxidative stress and inflammation as the central mechanisms of sulforaphane action. Preclinical studies have consistently demonstrated activation of the Nrf2 pathway, increased expression of downstream antioxidant enzymes (such as NQO1, GSTP, and GSH-related systems), and suppression of pro-inflammatory mediators (including IL-6, TNF-α, COX-2, and NF-κB). Clinical trials often focus on diseases in which oxidative stress and chronic inflammation play key pathogenic roles, such as neurodegenerative diseases, cardiovascular dysfunction, renal disease, and psychiatric disorders, and do not always directly measure these markers. The congruence of preclinical study objectives with clinical indications suggests a consistent biological rationale, even if direct biomarker validation in human studies is lacking.

The treatment of neurological and psychiatric disorders further illustrates the differences and complementarity between the tables. Clinical trials have examined sulforaphane in complex conditions, such as autism spectrum disorders, schizophrenia, psychosis, depression, and Alzheimer's disease, often focusing on behavioural outcomes, cognitive function, or symptom severity. In contrast, preclinical studies indirectly address neurological relevance through models of oxidative stress, mitochondrial dysfunction, autophagy, and protein aggregation in neuronal and non-neuronal cells. Although these studies provide plausible mechanisms of neuroprotection, they cannot replicate the complexities of human neuropsychiatric disorders. Therefore, the clinical table represents a crucial step beyond preclinical evidence, examining whether the cellular effects observed in experimental systems translate into significant functional improvements in patients.

Another significant difference was methodological heterogeneity. The table of preclinical studies is characterised by extreme heterogeneity in terms of experimental design, cell types, concentrations, exposure times, and endpoints measured. This heterogeneity reflects the exploratory nature of preclinical studies but also limits the direct comparability between studies. Clinical trials represent the most integrated level of evidence, focusing on whole-organism responses, therapeutic efficacy, and safety. Interestingly, while oncology studies dominate preclinical studies, human clinical trials have a somewhat different emphasis; as mentioned earlier, a significant portion of clinical trials focuses on cognitive, affective, and behavioural disorders.

This contrast highlights a common challenge in translational research: a wealth of preclinical research does not necessarily translate into consistent clinical results. In terms of research maturity, the preclinical studies table covers a broader time and conceptual scope, encompassing early proof-of-concept studies and more recent studies addressing specific molecular targets, such as ferroptosis, microRNAs, and epigenetic regulators. The clinical studies table shows that interest in sulforaphane has been focused on research since 2020. This temporal pattern suggests that decades of preclinical research have gradually accumulated sufficient justification for expanded clinical trials, although definitive late-phase studies remain limited.

Finally, the two tables differ in their epistemological roles in sulforaphane research. The preclinical studies table explains how sulforaphane works by identifying the targets, pathways, and cellular responses that can be experimentally manipulated. The clinical table discusses whether sulforaphane works in humans under real-world conditions in complex biological and social contexts.

The combined analysis of the tables does not provide redundant information but rather complementary perspectives on the topic. Preclinical data lend biological plausibility to clinical hypotheses, whereas clinical trials verify the validity and plausibility of these hypotheses (in practice). The convergence of findings regarding cancer, oxidative stress modulation, and anti-inflammatory mechanisms across both datasets supports the overall consistency of sulforaphane research, while discrepancies in scope, methodology, and results highlight areas where further integration of preclinical and clinical studies is necessary.

Despite the extensive evidence generated at the laboratory level, translating findings obtained "*at the benchside*" into effective "*at the bedside*" applications remains a significant challenge. Understanding this transformation requires careful consideration of the methodological and conceptual differences between experimental systems and clinical practices.

4.3 Formulation and Delivery Strategies

To exert therapeutic effects, sulforaphane must be efficiently delivered into the body, either as the active compound itself (SFN), its precursor glucoraphanin (GRA), or a combination of glucoraphanin with the enzyme myrosinase (GRA+MYR). The bioavailability of these forms varies significantly, with pure SFN exhibiting approximately seven times higher bioavailability than GRA and the GRA+MYR combination demonstrating approximately twice the bioavailability of GRA alone (Fahey et al. 2019). Zhou et al. (2018) estimated the half-life of SFN to be approximately 2 hours. However, Fahey et al. (2019) noted substantial interindividual variability in the metabolism of both SFN and glucoraphanin, highlighting the complexity of optimising SFN delivery for therapeutic purposes. However, its clinical utility is often restricted by its chemical instability and rapid metabolism. The use of nanocarriers, such as SFN-loaded nanoliposomes (SFN-NLPs) and polymeric nanoparticles, addresses these issues by enhancing stability, facilitating sustained release, and improving tumour-specific delivery (Alshammari et al. 2025; Joshi et al. 2025; Zhao et al. 2025). For example, Alshammari et al. (2025) noted that in Ehrlich Ascites Carcinoma models, SFN-loaded nanoliposomes (SFN-NLPs) significantly reduced tumour volume and prolonged survival more effectively than the free compound.

A major challenge in SFN applications is its inherent instability, as both SFN and myrosinase are highly susceptible to degradation (Yagishita et al. 2019; Yuanfeng et al. 2021). Several approaches have been investigated to enhance SFN stability and bioavailability. Among the most promising strategies are polyethylene glycol (PEG)-based ointments, which improve SFN solubility and dermal absorption; cyclodextrin inclusion complexes, which significantly enhance SFN thermostability; and micro-/nanoencapsulation techniques, which provide improved stability and controlled release (Wu et al. 2014; Rosenberg and Rosenberg 1944; Yagishita et al. 2019; Ali et al. 2023).

Initially, attempts were made to incorporate SFN into food matrices (Zambrano et al. 2019). In recent years, numerous studies have explored novel methods for SFN stabilisation (Table 4.2). These strategies aim to increase SFN bioavailability by modifying its polarity and enhancing its ability to cross the blood-brain barrier and preserve or even enhance its biological activity upon administration.

Table 4.2 Recent advancements in sulforaphane stabilisation methods

Material/patented-trade name	Role	Result	Author
Multi-walled carbon nanotubes (MWCNTs), chitosan (CS), polydopamine (PDA)	MWCNTs—drug carriers CS—base material PDA—bridge material	Enhance the stability and efficacy of SFN (drug loading was increasing with SFN concentration)	Li et al. (2024a)
Biologically inert alpha-cyclodextrin complex/SFX-01 (Sulforadex®)	Alpha-cyclodextrin in a 1:1 complex	Pharmacological activity was in the line with expectation (absorption in the health man)	Clack et al. (2025)
Solid lipid nanoparticles (SLNs)	SFN was encapsulated	Improved intestine bioavailability (in comparison with SFN) Not inhibited anticancer activity (A549 cancer cells)	Han et al. (2024)
Emulsome	Lipid-based SFN delivery system, SFN was encapsulated	Significant decrease was observed in proliferation of the cells (MCF7 breast cancer) Higher biological activity of free SFN (against MCF7)	Karroum and Üçışık (2024)

(continued)

Table 4.2 (continued)

Material/patented-trade name	Role	Result	Author
Be12O12 nanocage (inorganic) B12N12, Al12N12, Ga12N12 nanocages (nitrides)	SFN on nanocages with Be, B, Al, Ga	Theoretical beneficial system for delivery SFN	Wu et al. (2024a, b)
Silk fibroin protein nanoparticles (SFPNPs) conjugated with folic acid, polyethylene glycol	SFN capsulated	Release was pH-dependent (max -70 % – at pH = 7.4)	Ghanbari Hassan Kiadeh et al. (2024)
Nanocomplex: the amphiphilic peptides (C18GR7RGDS), a rose bengal (RB) and sulforaphane (SFN), /SFN@RB@SPM	Ultrasonic polymerisation of C18GR7RGDS encapsulated of sonosensitiser RB and SFN	SFN@RB@SPM crosses the blood-brain barrier and is easily taken up by U87-MG cells Induced much more apoptosis of U87-MG cells (compared to pure SFN)	Li et al. (2024b)
Microencapsulation with whey (BW) and pea protein (BP)	SFN freeze-drying microencapsulation with BW and BP	BW significantly improved bioaccessibility (67.7 ± 1.2 %) and bioavailability (54.4 ± 4.0 %) of SFN	Redha et al. (2025)
Ternary polymer hydrogel: chitosan, gelatine, polyethylene glycol	Gel with SFN nanoemulsion	Improves collagen production	Salehiamin et al. (2024)
Encapsulation in nanoparticles: propylene glycol alginate, zein, anthocyanin, $CaCl_2$	SFN in propylene glycol alginate with zein as wall materials, anthocyanin, and $CaCl_2$ as crosslinking agents	Slower release effect Improved bioavailability	Wang et al. (2024)
Nanoemulsions	SFN (half-life) in the nanoemulsion with high oleic sunflower oil	Higher bioavailability and lower oral clearance	Lee et al. (2023)
Complexed with α-cyclodextrin/SFX-01	A stabilised form of SFN	Inhibited proliferation of glioblastoma cell lines	Leung et al. (2021)
Nanoparticles/SBA-15	Adsorption of SFN based on SBA-15: silica functionalised with thiol and amine groups by the post-synthesis grafting procedure	Improves SFN absorption of the SBA-15 material	Hafezian et al. (2021)
Co-encapsulation nanoliposomes with basil seed gum (BSG)	Co-encapsulation of SFN in nanoliposomes and BSG	Decrease of the release speed Promoting release at pH 7.4	Azarashkan et al. (2022)

(continued)

Table 4.2 (continued)

Material/patented-trade name	Role	Result	Author
Microencapsulation: oil-in-water emulsion	SFN/Gum Arabic ratio of 0.82 and surfactant/oil ratio of 1.0	Improves thermal stability of SFN	Zambrano et al. (2023)
Plant nanocarriers	Membrane vesicles from broccoli to encapsulate SFN	Decrease in proliferation of SK-MEL-28 cells High activity for the inhibition of melanocyte development	Yepes-Molina and Carvajal (2021)
Free or encapsulated powder	Broccoli seed extract powder (BSE[GR]) with myrosinase-rich mustard seed powder ± ascorbic acid (a co-factor of Myr)	Capsules (72.1 %) were more efficient in conversion into SFN then free powder (29.3 %) Ascorbic acid increased the conversion level	Zhu et al. (2025)
Extracellular vesicles	SFN encapsulated into extracellular vesicles	Stimulation of hPC-12 cell proliferation Potentially: a gene expression regulator in autism spectrum disorder	Shahlaei et al. (2022)
Microencapsulation with chitosan, zein, and gelatine	SFN microencapsulated with different wall material	Improves thermal resistance Final release rate of SF was 80-90 %	Wang et al. (2021)
Nanoparticles	SFN with chitosan	Significantly suppress cell cycle arrest and activate apoptotic signals Medication effective delivery	Zhang et al. (2023)
Nanocarriers (M-HA-SS-TA)	Mineralised hyaluronic acid-SS-tetradecyl nanocarriers (M-HA-SS-TA) with SFN	Excellent tumour-targeting ability Tumour inhibition	Gu et al. (2021)
Nanoliposome with polyvinyl alcohol (PVA, 10 %)/ polyethylene glycol 400 (PEG400, 5%)	SFN encapsulated within the scaffold hydrogel system	Release tempo increased as the pH environments raise Possible wound-healing potential	Hemati et al. (2023)

Numerous material-based and nanotechnology-based strategies have been developed to protect SFN from degradation while maintaining or enhancing its biological activity. Among inorganic and polymeric delivery systems, multi-walled carbon nanotubes combined with chitosan and polydopamine have been proposed as efficient drug carriers, significantly increasing the SFN loading capacity and stability in a concentration-dependent manner (Li et al. 2024a). Similarly, the adsorption of SFN onto functionalised mesoporous silica (SBA-15) improved compound absorption through thiol- and amine-modified surfaces (Hafezian et al. 2021).

Cyclodextrin-based stabilisation represents one of the most advanced approaches, with the biologically inert α-cyclodextrin complex SFX-01 (Sulforadex®) forming a 1:1 inclusion complex that preserves pharmacological activity and demonstrates predictable absorption in healthy human participants (Clack et al. 2025). This formulation has also demonstrated antiproliferative effects in glioblastoma cell lines, supporting its translational potential (Leung et al. 2021). Lipid-based systems, including solid lipid nanoparticles, emulsomes, nanoemulsions, and nanoliposomes, have been widely explored. Encapsulation of SFN into solid lipid nanoparticles improved intestinal bioavailability without compromising its anticancer activity in lung adenocarcinoma cells (Han et al. 2024), whereas emulsome-based delivery systems reduced breast cancer cell proliferation, albeit with slightly lower biological activity than free SFN (Karroum and Üçışık 2024). Nanoemulsions formulated with high-oleic sunflower oil enhance SFN bioavailability and reduce oral clearance (Lee et al. 2023).

Protein- and polysaccharide-based encapsulation has also proven to be effective. Microencapsulation using whey and pea proteins significantly improved SFN bioaccessibility and bioavailability, with whey protein achieving bioaccessibility values exceeding 67 % (Redha et al. 2025). Chitosan-, zein-, gelatine-, and alginate-based micro- and nanoformulations enhanced thermal resistance and enabled controlled release, with final SFN release rates reaching 80–90 % under physiological conditions (Wang et al. 2021, 2024). pH-responsive delivery has been demonstrated using silk fibroin protein nanoparticles conjugated with folic acid and polyethylene glycol, achieving maximal release at physiological pH (Ghanbari Hassan Kiadeh et al. 2024), as well as hydrogels and nanoliposomes designed for wound healing and tissue regeneration (Hemati et al. 2023; Salehiamin et al. 2024).

Advanced multifunctional nanocomplexes have broadened the therapeutic scope of SFN. Amphiphilic peptide-based nanocomplexes co-encapsulating SFN and rose bengal were shown to cross the blood–brain barrier and induce significantly higher apoptosis in glioblastoma cells than free SFN (Li et al. 2024b). Tumour-targeted nanocarriers, such as mineralised hyaluronic acid–based systems, exhibit excellent tumour-homing ability and pronounced tumour inhibition (Gu et al. 2021). In parallel, inorganic nanocages composed of boron, aluminium, or gallium nitrides have been proposed as theoretical SFN delivery platforms with favourable binding properties (Wu et al. 2024a, b).

Biological carriers, including plant-derived membrane vesicles and extracellular vesicles, are a new class of delivery systems. Broccoli-derived nanovesicles effectively inhibit melanoma cell proliferation and melanocyte development (Yepes-Molina and Carvajal 2021), whereas SFN-loaded extracellular vesicles stimulate neuronal cell proliferation, suggesting potential applications beyond oncology, including in neurodevelopmental disorders (Shahlaei et al. 2022). Furthermore, formulation strategies that optimise the enzymatic conversion of glucoraphanin to

SFN, such as encapsulated broccoli seed extracts combined with myrosinase and ascorbic acid, significantly increase SFN yield compared to free-form powders (Zhu et al. 2025).

These stabilisation and delivery strategies demonstrate that formulation design can substantially enhance the physicochemical stability, bioavailability, and biological efficacy of sulforaphane. The diversity of materials and carrier systems highlights the versatility of SFN as a bioactive compound and supports its continued development for clinical, nutraceutical, and pharmaceutical applications.

4.4 Future Perspectives

Sulforaphane occupies an important position at the intersection of nutritional science and pharmacology. Its favourable safety profile and multi-targeted mechanism of action provide a solid rationale for further multidisciplinary research aimed at exploiting its full clinical potential as a therapeutically relevant compound.

Numerous studies have demonstrated the potential of SFN in treating a range of conditions, including cancer, cardiovascular diseases, and neurodegenerative diseases, which are broadly defined as inflammation-related disorders. Despite these promising findings, the clinical use of SFN remains limited because of its chemical instability and variable oral bioavailability. The conversion of glucoraphanin to active sulforaphane is highly dependent on both myrosinase activity and individual differences in gut microbiota composition, leading to inconsistent absorption and efficacy across patient populations. This represents a significant biochemical challenge, placing sulforaphane on the market as a dietary supplement rather than as an approved therapeutic agent. The main direction of future research will be the development of advanced drug delivery systems capable of improving the pharmacokinetic profile of SFN. Formulation strategies, such as nanoemulsions, microencapsulation, and cyclodextrin inclusion complexes, have shown great promise in improving the stability of compounds and enabling their specific delivery to specific tissues. Their further optimisation is likely to be a key area of interest in the coming years.

Another important direction is the stereochemical characterisation of SFN enantiomers. Given that R- and S-sulforaphane may differ in biological activity and receptor interactions, specialised studies with enantiomerically pure R-sulforaphane will clarify whether stereochemical specificity influences pharmacodynamic outcomes.

It also seems likely that future preclinical studies will need to expand beyond the (already explored) Nrf2-dependent response to encompass alternative signalling pathways involved in apoptosis, inflammation, and mitochondrial homeostasis. The integration of "omics" technologies, including transcriptomics, proteomics, and metabolomics, can provide an effective tool for elucidating the systemic mode of action of SFN and identifying molecular biomarkers that account for interindividual variability in treatment response.

It would also be beneficial to expand the database of human studies from those focusing on neurological and behavioural outcomes to those investigating oncological or cardiometabolic diseases.

Finally, investigating the synergistic interactions of SFN with conventional pharmacological agents is important when considering the growing database of drugs and their interactions with SFN (both beneficial and competitive).

Sulforaphane occupies an important position at the intersection of nutritional science and pharmacology. Its favourable safety profile and multi-targeted mechanism of action provide a solid rationale for further multidisciplinary research aimed at exploiting its full clinical potential as a therapeutically relevant compound.

References

Ali MA, Khan N, Kaleem N, Ahmad W, Alharethi SH, Alharbi B et al (2023) Anticancer properties of sulforaphane: molecular insights. Front Oncol 13:1168321

Alshammari AN, Alfawaz MS, Alenezi YM, Al-Ghafari AB, Al Doghaither HA, Elmorsy EM et al (2025) Nanoliposome-mediated sulforaphane delivery suppresses Ehrlich carcinoma. BMC Cancer 25:1838

Azarashkan Z, Motamedzadegan A, Ghorbani-HasanSaraei A, Rahaiee S, Biparva P (2022) Stability of sulforaphane-enriched nanoliposomes. Food Bioprocess Technol 15:1573–1587

Clack G, Moore C, Ruston L, Wilson D, Koch A, Webb D, Mallard N (2025) A phase 1 randomized, placebo-controlled study evaluating the safety, tolerability, and pharmacokinetics of enteric-coated stabilized sulforaphane (SFX-01) in male participants. Adv Ther 42(1):216–232

DrugBank (n.d.) Sulforaphane DB12422. https://go.drugbank.com/drugs/DB12422. Accessed 21 Jan 2025

Fahey JW, Wade KL, Stephenson KK, Panjwani AA, Liu H, Cornblatt G, Cheskin LJ (2019) Bioavailability of sulforaphane following ingestion of glucoraphanin-rich broccoli sprout and seed extracts with active myrosinase: a pilot study of the effects of proton pump inhibitor administration. Nutrients 11(7):1489

Fermini B, Coyne ST, Coyne KP (2018) Clinical trials in a dish: a perspective on the coming revolution in drug development. SLAS Discov 23(8):765–776

Ghanbari Hassan Kiadeh S, Rahaiee S, Azizi H, Govahi M (2024) The synthesis of broccoli sprout extract-loaded silk fibroin nanoparticles as efficient drug delivery vehicles: development and characterization. Pharm Dev Technol 29(4):359–370

Gu HF, Ren F, Mao XY, Du M (2021) Mineralized and GSH-responsive hyaluronic acid based nano-carriers for potentiating repressive effects of sulforaphane on breast cancer stem cells-like properties. Carbohydr Polym 269:118294

Hafezian SM, Biparva P, Bekhradnia A, Azizi SN (2021) Amine and thiol functionalization of SBA-15 nanoparticles for highly efficient adsorption of sulforaphane. Adv Powder Technol 32(3):779–790

Han L, Ma X, Chen M, He J, Zhang W (2024) Preparation, characterization and in vitro anticancer activity of sulforaphene-loaded solid lipid nanoparticles. Foods 13(23):3898

Hemati H, Haghiralsadat F, Hemati M, Sargazi G, Razi N (2023) Design and evaluation of liposomal sulforaphane-loaded polyvinyl alcohol/polyethylene glycol (PVA/PEG) hydrogels as a novel drug delivery system for wound healing. Gels 9(9):748

Joshi H, Rani I, Sharma V, Sharma U, Dimri T, Kumar M, Tuli HS (2025) Sulforaphane: a natural organosulfur having potential to modulate apoptosis and survival signalling in cancer. Discov Oncol 16(1):1–14

Karroum R, Üçışık MH (2024) Efficacy and safety of sulforaphane-loaded emulsomes as tested on MCF7 and MCF10A cells. Turk J Biochem 49(5):629–636
Lee TK, Hur G, Choi J, Ban C, Kim JY, Yang H, Kim JH (2023) Enhancing stability and bioavailability of sulforaphene in radish seed extracts using nanoemulsion made with high oleic sunflower oil. Food Sci Biotechnol 32(9):1269–1279
Leung E, Wright E, Baguley BC (2021) Effect of SFX-01 on proliferation of human glioblastoma cell lines. bioRxiv. https://doi.org/10.1101/2021.09
Li R, Hu X, Shan S, Li Y, Cui W, Liu L (2024a) Study on the controlled release properties of modified multi-walled carbon nanotubes on sulforaphane. Carbon Lett 34(2):757–765
Li Y, Yang X, Wei Z, Niu H, Wu L, Chen C, Fan H (2024b) Sulforaphane wrapped in self-assembled nanomicelle enhances the effect of sonodynamic therapy on glioma. Pharmaceutics 17(1):34
Pinlac RK, Comia LE, Epino GN, Fernandez RM, Madrid HS, Salvacion AS, Nas JS (2021) Benchmarking the physicochemical properties of 500 compounds for ADMET property prediction in *Caenorhabditis elegans*. Anim Biol Anim Husb 13(2)
PubChem (n.d.) Sulforaphane compound 5350. https://pubchem.ncbi.nlm.nih.gov/compound/5350. Accessed 21 Jan 2025
Redha AA, Torquati L, Bows JR, Gidley MJ, Cozzolino D (2025) Microencapsulation of broccoli sulforaphane using whey and pea protein: in vitro dynamic gastrointestinal digestion and intestinal absorption by Caco-2-HT29-MTX-E12 cells. Food Funct 16(1):71–86
Rosenberg M, Rosenberg Y (1944) Zhang J (2018) Microencapsulation of a model oil in wall system consisting of wheat proteins isolate (WHPI) and lactose. Appl Sci 8(10)
Salehiamin M, Ghoraishizadeh S, Habibpour A, Tafreshi S, Abolhasani MM, Shemiranykia Z, Esmaeili J (2024) Simultaneous usage of sulforaphane nanoemulsion and tannic acid in ternary chitosan/gelatin/PEG hydrogel for knee cartilage tissue engineering: in vitro and in vivo study. Int J Biol Macromol 271:132692
Shahlaei M, Saeidifar M, Zamanian A (2022) Sustained release of sulforaphane by bioactive extracellular vesicles for neuroprotective effect on chick model. J Biomed Mater Res B Appl Biomater 110(12):2636–2648
Veber DF, Johnson SR, Cheng HY, Smith BR, Ward KW, Kopple KD (2002) Molecular properties that influence the oral bioavailability of drug candidates. J Med Chem 45(12):2615–2623
Wang Q, Feng Z, Shi C, Lyu X, Fan D (2024) Sulforaphane nanoparticles coated with zein-propylene glycol alginate attenuate N-diethylnitrosamine-induced liver injury in mice. J Food Sci 89(10):6707–6719
Wang X, Meng G, Wang Q, Wang L, Fan D, Lü X (2021) Preparation and characterization of sulforaphane microcapsules by electrospraying. Food Sci 42(7):113–119. https://doi.org/10.7506/spkx1002-6630-20200404-04
Wu S, Li L, Liang Q, Gao H, Tang T, Tang Y (2024a) A DFT study of sulforaphane adsorbed on M12O12 (M = Be, Mg and Ca) nanocages. Mater Today Commun 38:107687
Wu S, Li L, Liang Q, Gao H, Tang T, Tang Y (2024b) A DFT study of sulforaphane adsorption on the group III nitrides (B12N12, Al12N12 and Ga12N12) nanocages. J Biomol Struct Dyn 42(23):12730–12741
Wu Y, Mao J, You Y, Liu S (2014) Study on degradation kinetics of sulforaphane in broccoli extract. Food Chem 155:235–239
Yagishita Y, Fahey JW, Dinkova-Kostova AT, Kensler TW (2019) Broccoli or sulforaphane: is it the source or dose that matters? Molecules 24(19):3593
Yepes-Molina L, Carvajal M (2021) Nanoencapsulation of sulforaphane in broccoli membrane vesicles and their in vitro antiproliferative activity. Pharm Biol 59(1):1488–1502
Yuanfeng W, Chengzhi L, Ligen Z, Juan S, Xinjie S, Yao Z, Jianwei M (2021) Approaches for enhancing the stability and formation of sulforaphane. Food Chem 345:128771
Zambrano V, Bustos R, Arozarena Y, Mahn A (2023) Optimization of a microencapsulation process using oil-in-water (O/W) emulsion to increase thermal stability of sulforaphane. Foods 12(20):3869

Zambrano V, Bustos R, Mahn A (2019) Insights about stabilization of sulforaphane through micro-encapsulation. Heliyon 5(11)
Zhang MQ, Wilkinson B (2007) Drug discovery beyond the 'rule-of-five'. Curr Opin Biotechnol 18(6):478–488
Zhang X, Peng X, Hui J, Li H (2023) Sulforaphane functionalized chitosan nanoparticles serve as an effective nanocomplex system for the treatment of human cervical cancer. J Cluster Sci 34(6):2963–2975
Zhao Z, Chen Q, Qiao X, Wang J, Ali ZTA, Li J, Yin L (2025) Sulforaphane in cancer precision medicine: from biosynthetic origins to multiscale mechanisms and clinical translation. Front Immunol 16:1702860
Zhou Y, Yang G, Tian H, Hu Y, Wu S, Geng Y, Wu W (2018) Sulforaphane metabolites cause apoptosis via microtubule disruption in cancer. Endocr Relat Cancer 25(3):255–268
Zhu W, Cremonini E, Mastaloudis AF, Mitchell AE, Bornhorst GM, Oteiza PI (2025) Optimization of sulforaphane bioavailability from a glucoraphanin-rich broccoli seed extract in a model of dynamic gastric digestion and absorption by Caco-2 cell monolayers. Food Funct 16(1):314–328

GPSR Compliance
The European Union's (EU) General Product Safety Regulation (GPSR) is a set of rules that requires consumer products to be safe and our obligations to ensure this.

If you have any concerns about our products, you can contact us on

ProductSafety@springernature.com

In case Publisher is established outside the EU, the EU authorized representative is:

Springer Nature Customer Service Center GmbH
Europaplatz 3
69115 Heidelberg, Germany

www.ingramcontent.com/pod-product-compliance
Ingram Content Group UK Ltd.
Pitfield, Milton Keynes, MK11 3LW, UK
UKHW022107180926
475695UK00001B/32
9783032343307